EATING CLEAN²

Overcoming Food Hazards

A CONSUMER'S GUIDEBOOK

Selected Readings

INTRODUCTION BY
Ralph Nader

PREPARED WITH THE ASSISTANCE OF
Katherine Isaac and Steven Gold

DESIGNED BY
Kathleen Glynn

Center for Study of Responsive Law
Box 19367
Washington, D.C. 20036

ISBN Number 0-936758-21-X

Library of Congress Catalog Card Number 87-73555

TABLE OF CONTENTS

INTRODUCTION

For the Safety of the Food Supply - You're The Difference

BY RALPH NADER

Food safety is very much on people's minds – judging by recent polls and the letters coming into government agencies and consumer groups. Foodborne illnesses and disease are worrying people as they read about sudden illness epidemics. For example in the past two years, headlines reported over 16,000 consumers in Illinois fell ill due to salmonellosis from contaminated milk sold by a major supermarket chain. On the West Coast, 29 persons died and many more became sick from listeriosis after they ate Mexican-style cheese. Over 750 New Yorkers became sick after eating polluted raw clams. Hundreds of Californians were stricken after they consumed watermelons contaminated with the chemical, Aldicarb. The victims list continues with botulism from dried salted whitefish, viral hepatitis A from contaminated food, staphylococcus from tainted ice cream, cholera from bad seafood, gastroenteritis outbreaks on Caribbean cruise ships, summer camps and nursing homes.

These are the more publicized symptoms of a deeper, more widespread imperiling of the foods Americans eat. Thousands of pesticides, additives, antibiotics and other chemicals are used in the growing, beautifying, storing and preparation of foodstuffs. Not all are benign. Millions of tons of pollutants are pouring into lakes, bays and estuaries. Some of their residues reach our table via seafood, especially the shellfish. The struggle against filth and unsanitary conditions in meat and poultry processing plants, warehouses and retail stores needs to intensify as Reagan's deregulation of food safety programs weakens federal inspection and undermines law enforcement against wrongdoers.

Also, the U.S. is importing more food than ever before, including large quantities of fruits, vegetables, meats, and canned ethnic foods from countries whose inspection standards are even weaker than those Washington presently observes.

Water for cooking comes through water works whose purification systems are both antiquated and overwhelmed by polluted surface and groundwater supplies.

As if all this is not enough concern, now come companies pressing to irradiate the food of millions of Americans by exaggerating the benefits and ignoring the diverse risks both to workers and consumers.

But, coming rapidly over the horizon are ways to reverse this dismal slide into self-destructive ruts of chemicalized agribusiness and its far-reaching circles of poisons. The organic farm movement is safer and produces more nutritious and flavorful food. Recently, more and more farmers who are converting to integrated pest management modes are showing lower costs and equivalent or higher yields per acre. Moreover, they are off the chemical treadmill caused by faster mutational resistance of bacteria and insects.

More consumers are responding to the advantages of organically grown food. The trend is unmistakably toward fresh fruits and vegetables and against their canned counterparts. It is like a good genie coming out of the bottle. Soon a critical momentum of buyer knowledge and action may appear both in the marketplace and on local, state and federal governmental agencies which are supposed to protect our food supply. Getting to those happier objectives is the purpose of this Reader and Guidebook.

If you ask "What can I do?" this volume responds: "Plenty." Immediately, you can acquire the information to buy safer and more nutritious food with less fat, salt and sugar. You can demand to know when sulfites are used to fake freshness. Immediately you can save money in learning what not to buy as well as what to buy. For more than a few people, growing more of your own food, making more of your baby food and regular meals, instead of relying on the more expensive processed "convenience" stuff, can make sense right away. Strengthening consumer groups can make the government regulators for food safety reflect not industry's, but your demands to clean up the food chain. The material in this collection, together with free publications and other more specialized resources listed at the end, can open up practical possibilities in your kitchen, your grocery market and in agriculture to defend the integrity of the nation's foodstuffs.

A "can-do" attitude and an enthusiasm for working with other citizens in a common quest for "eating clean" are the prescriptions for success. Just being part of this national movement – spearheaded by such groups as "Americans for Safe Food" and others described in section Five – will sharpen your food buying and civic skills. Most important in the long run, organized consumer efforts for the public health set an enduring example for children – and do they ever need improved food habits to ward off the thousands of manipulating TV ads for junk food!

If you want to be on citizen alert, simply write to Americans for Safe Food at 1501 16th Street, NW, Washington, DC 20036 and ask them to keep you informed. You'll find their reports, releases and campaigns exciting – because they are important and directed toward enlisting grass roots support from Americans who want to band together for corrective action.

Let us begin!

SECTION 1
Eater Alerts to Major Problem Areas

Even with incomplete reporting, the federal government estimates 9,000 fatalities and at least 80 million foodborne illnesses a year in the United States. From wrenching stomach upsets to diarrhea to serious bacterial food poisoning, this toll produces pain, medical and hospital bills, and lost time from work. Yet, curiously, such a large wave of afflictions rarely leads to systematic correction or prevention.

This section is designed to motivate such action. The Center for Science in the Public Interest's *Nutrition Action* provides an overview in "Germ Wars: Cleaning Up Our Food." Other articles report on the mounting controversy over contaminated food and what can be done by training and law enforcement toward curbing unsanitary conditions in restaurants and other retail outlets. Does your daily newspaper print the health code violations by restaurants as released by the local or state department of health? To illustrate, the list for one week in New York City is printed here.

Next come eyebrow-lifting descriptions of the toxic chemicals added to our food. This chemical cuisine is an accumulating form of silent violence in the human body. It is adept at hiding its tracks, even when the person succumbs to cancer or some other disease. Birth defects are even harder to trace causally. Americans for Safe Food lists some of the toxic herbicides, pesticides and fungicides, by way of background, before taking readers on a journey through the miasma of pesticide permeated foods, afflicted farmworkers, and the farce that masks itself as safety regulation.

Then *Harrowsmith* magazine features "The Freshness Illusion" which, among other revelations, takes you down to a Mexican valley where many of your winter vegetables are born and raised in chemical baths and sprays. Follow the trail to your dinner table and you'll know how the "freshness" is squeezed from our edibles in more ways than one. But, there is a stirring among the people to find a way around these chemicals or at least for the time being, reduce their presence on the purchased food. Carol Sugarman of the *Washington Post* describes these marketplace responses. Then the story of Alar, the cosmetic chemical, and how, after the Reagan administration punted, consumer protest drove it out of most apple products.

The price of perfectly "appearing" produce is an additive avalanche inundating the diet. The Food and Drug Administration's magazine *FDA Consumer* gives you a primer on artificial colors, flavors, sweeteners, emulsifiers (mixers) and stabilizers, among others. The appeal to the senses, from the tongue to the eye, keeps many a chemical factory busy. The strategy of turning a consumer's senses against his or her common sense keeps many a Madison Avenue advertising agency busy as well. Removing dangerous additives from your food supply has been a battle that would stretch anyone's patience. The *Health Letter* shows what forces came into this battle from the government, industry and consumer sides. Illustrating how a new offensive is launched, the Center for Science in the Public Interest announces in late 1987 its drive against cancer-causing urethane in alcoholic beverages. A recent pro-consumer court victory against the FDA regarding risky color additives is noted. Additives can also come from lead in cans and plastic in packaging of foods, caution the Goldbecks in their latest guide to good food.

Next is a truly ominous backfire—namely, the heavy overuse of antibiotics in rearing cattle, pigs and chickens which, *Nutrition Action* points out, is "breeding strains of drug-resistant microorganisms that make these medications worthless in fighting human disease."

Not even an Army barracks full of recruits can get messier faster than meat and poultry slaughter and packing plants. Many federal meat and poultry inspectors are worried about too much dirty, drugged and diseased products getting through to market. They decry the lack of support and resources from a deregulation-minded Washington. The problems are noted particularly in a letter from Americans for Safe Food to the Chairman of the House Subcommittee on Livestock, Dairy and Poultry. Our Center's report "Return to the Jungle" (See Resources Section) gave voice to the specific concerns of federal inspectors in 1982. More attention and backing should be given to these conscientious "guardians of our food supply."

At least there is a legal framework for blocking bad meat, unlike seafood which reaches your plate without any required federal inspection. We lost a battle narrowly in the Sixties when the fish industry would not go along with continual inspection legislation and Congress did not override. So, read what Public Voice reports about hazardous fish and what needs to be done. Here is an opportunity, while fish consumption is rising, to convince Congress and the fish industry that safety regulation is in everyone's best interest.

Americans now know about contaminated drinking water and want action. *Buyer's Market* has prepared a series of questions for you to send to your community water supplier, who should respond to them. A Public Citizen feature provides a broader background. We need to make a good existing federal law work under a reluctant administration or by a more responsive successor. The EPA finally is beginning to move on this front since Congress required deadlines for controlling pollutants in a 1986 amendment to the Safe Drinking Water Act.

A major component in our nation's overall foreign trade deficit is a surge of imported foodstuffs whose dollar value almost equalled our exports in 1986. Inspection of food imports is very erratic, almost casual. On March 26, 1987 the conservative *Wall Street Journal* printed a long, page one story titled "Poison Produce: As Food Imports Rise Consumers Face Peril from Use of Pesticides." Many fruits, vegetables and coffee tonnage enter our country laden with pesticide residues long banned by Washington for domestic agricultural applications but permitted for export. Thus, the book titled *The Circle of Poison*, which is excerpted here. Note the horrific exposure of foreign farm workers to the profligate misuse of these toxic pesticides. Note also how it took an environmental organization, the Natural Resources Defense Council (NRDC), to conduct its own state of the art tests to show Washington the large amount of contaminated coffee. NRDC has a few tips on how to minimize pesticides in your food, while advocating for a more basic prevention policy.

Many Americans think all their regular fruits and vegetables come from this country. Whether for apple juice, orange juice and many vegetables, this assumption is less and less true. Your supermarket manager should be able to tell you the country of origin for each product. But you have to ask. Canned or packaged imported foods are usually labeled as such, but *FDA Consumer's* article herein tells why buyers need to be cautious toward this "exotic ethnic fare."

And concluding this section is some sobering information about irradiated food which, following FDA approval, some companies are pushing to make routine in coming years.

Germ Wars:Cleaning Up Our Food
BY ELAINE BLUME

NUTRITION ACTION HEALTHLETTER, JUNE, 1986, P. 1+.

Suddenly, you don't feel quite right. You're aware of a queasy sensation in the pit of your stomach, but since you were fine just a moment ago, you hope it's only your imagination. Soon, though, you feel weak and ill; then vomiting and diarrhea begin. You wonder whether you got food poisoning from that dreadful cafeteria at work, or if it's just a bug that's going around.

Both are plausible theories. Diarrheal illness is extremely common; many different microbes can cause it, and all of them can be found in our food. Relevant statistics are sketchy, but experts say the problem is getting worse.

With more elderly and otherwise vulnerable people in the population, and with bacteria becoming increasingly resistant to life-saving antibiotics, what starts out as an uncomfortable bout of diarrhea may now more often become a serious threat to health. New evidence also suggests that foodborne infections may precipitate disabling conditions such as arthritis. Meanwhile, the government continues to use an antiquated system to inspect meat and poultry and does not inspect most seafood at all.

Between 69 million and 275 million episodes of diarrhea occur in the United States each year – as many as one per person on average[1], estimate Food and Drug Administration (FDA) microbiologists Douglas L. Archer and John E. Kvenber. Though relatively few of these cases are ever traced to their source, the scientists reckon that contaminated food accounts for about one-third of the total.

"Microbial contamination of food is the most important health problem that still afflicts the food supply," says Thomas P. Grumbly, formerly an official both in the Department of Agriculture (USDA) and in the FDA. "Government isn't adequately addressing this issue, because they feel that the consumer can deal with the problem, and because they don't really know how to clean up the food without leaving it a tasteless mess. It's a serious health problem – disguised by the fact that everyone gets tummy aches from time to time."

Tracking Tummy Aches

Where do these "tummy aches" come from? The answer is hard to pin down, because available statistics are so imprecise. An outbreak of foodborne illness is much more likely to be reported if it's large, traceable to a restaurant, or leads to serious illnesses or deaths, than if it involves just a few cases of relatively mild sickness following a family cookout. And those illnesses with short incubation periods, like food poisoning caused by *Staphylococcus* bacteria, are more likely to be recognized as foodborne than illnesses like hepatitis, which may take weeks to show up.

Scientists point an accusing finger at raw poultry, meat, fish, eggs, and dairy products, which are known to harbor most of the harmful microbes in our food supply. In 1982, about 23 percent of the illness outbreaks that could be traced to a food source came from meat and poultry; 37 percent were blamed on fish and shellfish.[2] In this country, fruits and vegetables rarely cause problems.[3]

Bacteria – microscopic one-celled organisms that can be killed by antibiotics – are the culprits in about two-thirds of foodborne illness outbreaks. Viruses – which are smaller an unaffected by antibiotics – cause about another 10 percent. Most of the remainder involve poisoning from chemical contamination.[2]

For most people, "food poisoning" is usually very unpleasant but not terribly serious. However, for infants, fetuses, the elderly, and those suffering from other illnesses, it can be fatal. Scientists from the Centers for Disease Control have estimated that foodborne infections kill some 9,000 persons in this country every year.[4] The economic impact of foodborne illness is great, too. Medical costs, lost wages, and recalled and destroyed food amount to $1 billion to $10 billion each year, according to microbiologist and medical economist E.C.D. Todd.[5]

Crippling Legacy

But immediate illness, death, and economic loss are just part of the story. In a small percentage of patients, diarrheal infection may leave the disabling legacy of a serious chronic ailment, such as arthritis, kidney damage, or heart problems. Victims may also suffer short term or long-term intestinal damage that interferes with their ability to absorb nutrients, leaving them weak and vulnerable to other illness.[1]

Diarrheal infection seems to work together with genetics to produce certain distinct but probably related forms of arthritis, including ankylosing spondylitis, Reiter's syndrome, and reactive arthritis. Sometimes, these illnesses lead to heart damage, as well as crippling inflammation of the joints. Sixty to 80 percent of patients with these forms of arthritis may have cells with a marker known as HLA-B27, which is present in only about eight percent of the population at large. (Markers like HLA-B27, which are present on many different types of cells, are analogous to the A,B, and O markers which are found on red blood cells and determine blood type. Both kinds of cell markers are inherited.)

However, the HLA-B27 marker isn't a sure ticket to these crippling diseases. Even in a pair of identical twins with the marker, one of the pair may develop arthritis while the other does not.[6] And since these special forms of arthritis often follow on the heels of a diarrheal infection, researchers believe the infection may trigger the arthritis. If that's true, we might be able to decrease the incidence of these chronic diseases if we can succeed in lowering the incidence of diarrhea.

Studies also suggest that diarrheal infection may sometimes be followed by kidney failure, immune deficiency, allergy, or rheumatoid arthritis. More often, diarrhea may lead to continued difficulty digesting and absorbing food; this is especially likely to happen in children. FDA's Archer and Kvenberg warn that nutritional loss from diarrhea and its complications "may greatly increase morbidity from other causes and generally erodes the overall health of the population."

The Culprits. . .

Of the invisible microbes that cause all this damage, Salmonella ranks as a major culprit. It causes diarrhea ranging from mild to lethal. Farm animals and poultry carry these bacteria in their intestines, and when infected livestock come to market Salmonella comes too. Proper kitchen hygiene circumvents the problem, but few of us are perfect.

"*Salmonella* infections are on the rise," says Carol Tucker Foreman, the former assistant secretary of agriculture and long- time consumer advocate, due to "drug-resistant *Salmonella* strains and an increased number of immunosuppressed people around."

Until recently, scientists thought *Salmonella* was the principal disease-causing microbe in meat and poultry. But surveys now suggest that the newly recognized bacterium, *Campylobacter*, is even more common.[7] Like *Salmonella, Campylobacter* causes sometimes severe but rarely fatal diarrhea in humans. And many strains of both *Salmonella* and *Campylobacter* are now antibiotic-resistant, making them even more of a threat.[7,8]

Antibiotics are used in animals and poultry at high levels to treat illness and at low levels to enhance growth. Currently, animal use accounts for about half of the more than 35 million pounds of antibiotics produced in the United States. And scientists have shown that resistant strains of microbes may develop in animals receiving antibiotics and can then be passed to humans (See *Nutrition Action Healthletter*, May 1986).

This sequence of events increases the danger that our life- saving antibiotics may become less effective, and makes it imperative to restrict the use of antibiotics in animals and to limit the spread of bacteria from animals to humans.

Who's Responsible?

Meanwhile the bacteria *do* spread, and anyone who has spent a day limping from bed to bathroom and back again has probably wondered whom to blame. While FDA is responsible for ensuring the overall safety of our food supply, USDA has jurisdiction over meat and poultry, and its Food Safety and Inspection Service runs this country's massive meat and poultry inspection program. But USDA doesn't reject meat or poultry simply because dangerous microbes are present: rather it takes the position that such contamination is inevitable.[9]

Others disagree. "The law says the government shouldn't stamp as 'approved' anything that contains a contaminant that might be a hazard to health," says Foreman. "In my opinion, they're breaking the law every day."

Cleaning Up Contamination

In response to the growing problem and public criticism, USDA and industry have been developing measures to reduce *Salmonella* contamination, particularly in chickens. Both are sponsoring research aimed at producing cleaner chicken feed, keeping breeder flocks *Salmonella*-free, and colonizing chickens' intestinal tracts with harmless bacteria able to "competitively exclude" *Salmonella*.

Trouble spots in meat and poultry processing have also attracted attention. Defeathering birds is a remarkably unsanitary process, points out Dr. Morris Potter, a veterinary epidemiologist at the Atlanta-based Centers for Disease Control. During this procedure, the machinery holding and plucking the chickens presses bacteria-laden feces out the freshly killed birds, soiling the "rubber fingers" that pluck the feathers.

As they work, these "fingers" then beat the feces into the chickens' skin, and especially into the empty follicles that once held the feathers. Right now, most plucking machinery is difficult to clean. It could be redesigned to solve this problem, and cleaning methods could be improved as well.[10]

After defeathering and evisceration, the still-warm birds are plunged into a "chiller bath" of icy water. This bath, too, can spread bacteria from one chicken to another.[11] Adding chlorine to the chiller water interferes with such transfer, but experts now question whether it's safe to use chlorine in food processing.[12]

It may be best to avoid the chiller bath altogether; in Sweden and some other countries, the chickens are cooled by spinning instead.[13]

The inspection of meat and poultry also could be improved. In 1983, USDA's Food Safety and Inspection Service (FSIS) asked the Food and Nutrition Board (FNB) of the National Academy of Sciences to evaluate FSIS' inspection system; FNB issued its report in 1985.[14]

FNB concluded that it was impossible to know how many infections the inspection program has prevented or whether FSIS' intensive program of poultry inspection "protects consumers against any diseases of public health importance." Pointing out that current inspection methods were inadequate for detecting *Salmonella* or *Campylobacter*, FNB urged FSIS to intensify its efforts to rid meat and poultry of these microbes. The board also recommended that:

- new inspection methods be objectively evaluated on the basis of their impact on health;
- USDA adopt a trace-back and recall system, so that problems could be traced to their source and dealt with quickly;
- inspectors use rapid, inexpensive screening tests to detect hazardous microbes; and
- inspectors make surprise visits to plants in which inspection is not continuous.

The Food and Nutrition Board also cited the need for improved testing for pesticides and other toxic chemicals.

Free-Market Fish

While meat and poultry inspection may need improvement, fish and shellfish inspection barely exists. This is ironic considering that virus-contaminated shellfish have caused many large outbreaks of illness in recent years, and seafood in general is responsible for more than its fair share of food poisoning.

Current law does not require inspection of fish and shellfish or of the plants in which they are processed. The Department of Commerce's National Marine Fisheries Service inspects fish only when a manufacturer requests, and pays for, the inspection. (Usually, a manufacturer makes such a request for promotional purposes, or in order to sell the products to an organization that requires certification.) States carry out some non-voluntary inspection. But the majority of fish and fish products comes to the supermarket without any inspection oversight at all.[15]

According to Public Voice for Food and Health Policy, a Washington-based consumer group, "the frequency and intensity of [fish] inspection is wholly inadequate for the nature of the product."[16] Both germs and toxic chemicals can pose some real problems in fish and even some industry leaders agree on the need for a certain amount of inspection to maintain consumer confidence.

But just how this inspection should be conducted is a matter of debate. Effective consumer protection would require more than a quick check for smelly, decomposing fish. It would entail rapid microbiological assays, tests for toxic chemicals, and assurance that the fishing grounds were not contaminated. Otherwise, mandatory fish inspection might turn into an expensive exercise reaping few health benefits.

"USDA Method Spreads Germs."

Budget contraints are pushing the Department of Agriculture to modify its own continuous inspection program for meat and poultry. A legacy of *The Jungle*, Upton Sinclair's 1906 expose of the filth-ridden meatpacking industry of that era, the present system requires that each individual animal or bird be inspected at the time of slaughter. Many of the problems uncovered by this intensive inspection are defects, such as bruises on chickens, that impair quality but have no bearing on human health.

"Looking at every chicken is a waste of money designed for another era," says Grumbly. "Government inspectors touch one bird after the other and actually *spread* infection." USDA itself would like to move to a sampling system for poultry inspection, but this change would require new legislation.

The department also would like to customize inspection, so it can direct resources where they are most needed. For example, it might spend more time watching small, marginal plants than massive operations with easily recognized trade names. The latter are eager to protect their reputations and generally have their own effective systems of safety and quality control.

Better inspection and new technologies should lessen problems of microbial contamination but right now the microbes are gaining on us. And many believe too little is being done to reverse this trend. As dirty food collides with Gramm-Rudmanized budgets, it's all too likely that the germs will prevail. ◆

Elaine Blume, an experienced science and health writer, recently joined CSPI's staff.

Footnotes

[1] J. Food Protection 48: 887, 1985.
[2] Centers for Disease Control. Foodborne Disease Surveillance: Annual Summary, 1982 (issued September 1985), pp.8-9.
[3] National Research Council. An Evaluation of the Role of Microbiolgical Criteria for Foods and Food Ingredients, 1985, p. 257.
[4] Carter Center. Closing the Gap, 1986.
[5] U.S. Food and Drug Administration, Contract No. 223-84-2087. Proceedings of the Second National Conference for Food Protection, 1984.
[6] J. Food Protection 48: 538 (1985).
[7] Am. J. Pub. Health 76: 401, 1986.
[8] Proceedings of the International Symposium on Salmonella, Am. Asso. of Avian Pathologists, 1984, p. 17.
[9] USDA; FSIS-9. Food-Borne Bacterial Poisoning, 1980.
[10] Proceedings, pp. 252-253.
[11] Proceedings, p. 251.
[12] National Research Council, p. 63.
[13] Thomas P. Grumbly, telephone interview.
[14] National Research Council, Meat and Poultry Inspection: The Scientific Basis of the Nation's Program, 1985.
[15] Paul Comar, National Marine Fisheries Service, telephone interview.
[16] Public Voice, A Market Basket of Food Hazards: Critical Gaps in Government Protection, 1983, p. 14.

The Diarrhea You Prevent May Be Your Own

It's tempting to blame that greasy luncheonette for your upset stomach, but the truth is, many cases of food poisoning get started at home. Since it's unlikely that the city health department will force the home cook to clean up his or her act, these sanitation tips – compiled by Elaine Blume – will help keep kitchen microbes under control.

Wash your hands before preparing food.

Assume that raw meat, fish, poultry, and dairy products are contaminated. Keep them refrigerated, and treat them with the respect they deserve. You may decide that steak tartare, oysters on the half-shell, or salad coated with raw egg is worth the risk, but realize that some risk does exist.

To be made safe to eat, raw foods need to be adequately cooked, pasteurized, or sterilized. Cook meat and poultry completely, without interruption; partial cooking encourages growth of bacteria. Even some cooked foods, such as steamed clams and rare roast beef, may harbor microbes unless they have been heated sufficiently.

Thorough cooking won't help if the food becomes contaminated after it is cooked. This is likely to happen if microbe-bearing raw foods are not kept sufficiently separate. Before they touch anything else, wash dishes, hands, and implements that have come into contact with raw meat, fish, poultry, or dairy products. Kitchen sponges and faucet handles are other potentially important sources of cross-contamination. And because they can harbor bacteria, wooden cutting boards are not recommended. Hard plastic boards are dishwasher-safe and don't warp when they get wet, as wood does.

Ideally, you should clean the whole work area used for raw meats before using it for other food; better yet, try to maintain separate work areas for raw and cooked foods.

The proper temperature for a refrigerator is between 34 and 40 degrees F., and for a freezer, zero degrees F. or below.

Keep food cold or hot, not in between. ("Cold" in this context is 40 degrees F. or below, and "hot" is 140 degrees F. or above.) In general, perishable foods shouldn't remain at or near room temperature for more than two hours. At 60 degrees F., illness-producing bacteria can start to grow: at 80 degrees F. and above, they multiply very rapidly.

If you are dealing with a large quantity of food, consider dividing it into smaller portions for heating, serving (as at a buffet), or storage. This helps ensure that the food heats or chills through and also that it has enough exposure to air to discourage *Clostridium perfringens*, the "cafeteria germ."

Thaw food in the refrigerator, not at room temperature. Bacteria can start multiplying on the outside even while the center portion is still frozen.

Stuffed poultry presents special hazards. Because poultry is often contaminated, bacteria from the bird can get into the stuffing, which does not always get heated enough to kill the microbes.

Don't stuff poultry until it's ready to go in the oven, and pack the stuffing loosely, so it can get heated throughout. After the meal, remove leftover stuffing from the bird, and refrigerate it separately.

Remember that each misstep with food makes any additional misstep more risky. For example, sterile or near-sterile food can probably stand outside of the refrigerator safely for several hours, as long as it is then cooked thoroughly, or consumed promptly. But if the food contains some harmful bacteria to start with, they can reach dangerous levels very quickly in a warm room.

Use special care when preparing a brown-bag lunch or picnic. Keep perishable items refrigerated until you leave the house, and, if possible, refrigerate again when you arrive at work or school. If you can't refrigerate your lunch, keep it in the coolest place possible. Also, try putting something cold in the bag, like a frozen container of juice, or your sandwich, frozen overnight. And use thermoses to keep hot things hot and cold things cold. For a picnic, use a cooler with ice or frozen gel.

Don't use food from such damaged or suspect containers as bulging or dented cans; jars with cracks or bulging lids; cans that spurt liquid when opened; or paper packages with leaks or stains.

And don't check suspicious food by tasting it. It is an unfortunate – and dangerous – fact that microbes that make people ill usually do not cause food to spoil. And if a food contains botulinum toxin, eating even a small amount may prove fatal. So, when in doubt, throw it out! If you eat home-canned foods, learn and take proper precautions.

The Department of Agriculture maintains a toll-free hotline that will answer questions about proper handling of meat and poultry products. The number is (800) 535-4555 or in the Washington, D.C. area, 447-3333. District offices of the Food and Drug Administration will also answer questions about food safety. ◆

The New Consumer Crusade:

Momentum Builds in the Fight Against Contaminated Food

BY PATRICIA PICCONE MITCHELL
THE WASHINGTON POST, SEPTEMBER 16, 1987, P. E1.

Food poisoning is nothing new – the little bugs that cause people to get sick have been around for a long time. What is new is a growing awareness among consumers about contaminated food –and a growing demand that something be done about it.

"There's an unusual amount of momentum"

among consumers concerning the issue of food-borne illness, says Deborah Schechter, director of Americans for Safe Food, a project of the Center for Science in the Public Interest, a Washington advocacy group. And there are signs the problem is beginning to capture the attention of legislators, the industry and government officials.

For example, a bill expected to be introduced shortly by Sen. Patrick Leahy (D-Vt.) chairman of the Senate Agriculture Committee, would make some drastic changes in the way the Department of Agriculture inspects meat and poultry.

According to a draft version of the bill, USDA would be required to conduct a monitoring program to detect disease- causing microorganisms in meat and poultry products and to set limits on the number of organisms that such products could contain.

Under such a system, only a small number of the birds produced by a poultry plant could be contaminated with salmonella or other disease-causing organisms. USDA data shows that approximately one in three broilers produced is contaminated with salmonella. The department has never released contamination- rate statistics on an individual plant basis, but it is known that some plants maintain consistently lower rates of contaminated poultry than others.

Plants selling products that violate microbiological limits could eventually be put out of business under the bill currently being discussed. A similar monitoring system, to be operated by USDA, would be established for fish and shellfish that exceed standards for microorganisms and toxic chemicals.

"It's a really good first step," says Diane Heiman, government affairs director for Public Voice for Food and Health Policy, a Washington-based consumer group that has actively campaigned for changes in the government's approach to food safety. Last year, Public Voice published a report documenting the hazards related to tainted fish and shellfish and calling for a mandatory fish inspection program.

While mandatory inspection programs operated by USDA exist for meat and poultry products, no such program has ever been created for fish products. The Commerce Department's National Marine Fisheries Service (NMFS) runs a voluntary program, which some companies utilize in their marketing campaigns as an indication of quality.

Setting legal limits on microorganisms for raw meat and poultry "would be premature," according to Mahlon Burnette, technical consultant for the National Broiler Council, a trade association for chicken processors.

Burnette says there is no data base to use for setting acceptable levels and no known correlation between levels of microorganisms and human illness. Several National Academy of Sciences reports over the last few years have rejected the idea, he says, including an NAS report issued last May on poultry inspection.

USDA "doesn't have the capability of setting tolerances," agrees Susan Magaw, director of congressional relations for the American Meat Institute. Under the draft bill, companies that can't meet the

tolerance would be "out of business," notes Magaw, adding, "That's pretty tough."

However, Schechter argues that tolerances would force companies to make needed changes in their production systems. Heiman says companies would have an incentive to conduct their own testing to make sure their products meet the legal limits.

A research funding bill that has the support of industry groups and USDA was introduced recently by Sens. David Pryor (D-Ark.), Christopher Bond (R-Mo.) and Ted Cochran (R- Miss.). Rather than imposing any new regulatory system, the legislation would create a two year, $45 million research program to determine where contamination occurs in the meat and poultry food chain and how it can be prevented.

USDA officials have always asserted that the problem of contaminated food lies not just with the processor, but spans the entire system, beginning with the feed given to livestock and poultry ending with sloppy food handling practiced by food service personnel and consumers. William H. Dubbert, assistant deputy administrator for science in USDA's Food Safety and Inspection Service (FSIS), told a recent symposium on salmonella at Texas A&M University Research and Extension Center at Dallas that "we can't deal with the salmonella problem by addressing only a single phase of the cycle."

USDA appears bent on sticking to its current policies. Nancy Robinson, director of information and legislation at FSIS, says beyond ongoing programs like research projects and development of better testing methods to detect organisms, the agency has no plans for any new major initiatives such as those suggested by Leahy.

USDA recently turned down a request by a coalition of consumer groups to require all fresh poultry and meat to be labeled with cooking and handling instructions. However, Burnette says consumers will begin seeing poultry labels with handling and cooking information in the "very near future," as part of a voluntary labeling program recently adopted by NBC's board of directors. Widespread participation among companies is anticipated, he says. In addition, both NBC and AMI are developing recommended manufacturing procedures, to be distributed to companies for their voluntary adoption.

The fish industry has traditionally favored a strengthened inspection system as a way to increase product quality across the industry, says Dick Gutting, vice president for government relations at the National Fisheries Institute, a trade association of harvesting, processing and marketing companies.

Last year, Congress ordered NMFS to conduct a two-year study of a mandatory seafood surveillance system. The agency recently announced that it will contract with the National Academy of Sciences to focus on the health and safety issues involving seafood.

According to Public Voice, in 1982 seafood was responsible for 24 percent of food poisoning outbreaks for which the food containing the harmful agent could be identified. Meanwhile, fish consump-

tion is rising, and in 1985 reached a record 14.5 pounds per capita, an 11 percent increase since 1980. However, the study may take as long as three to four years to complete, and Gutting says the industry would probably want to see the study completed first before a surveillance or inspection program is implemented. ◆

Putting retail food safety to the test

BY CHRIS LECOS

FDA CONSUMER, MAY 1985, PP. 21-23.

The owner of the small restaurant glances at the clock and sees that he has only two hours to get everything done before opening. There's chicken to be baked, salads to be made, hamburgers to be readied for the always popular "Burger Special," and a large pot of chili, made the night before, to be reheated.

He takes a dozen chickens from the walk-in refrigerator, portions them into serving sizes on a cutting board, and whisks them into a preheated oven. He takes a cloth, wipes the knife and board, and then chops and mixes the salad ingredients before hand-filling equal amounts into wooden bowls.

He then removes 20 pounds of ground beef from the refrigerator, adds salt, pepper and a touch of garlic powder, mixes it well, and starts packing it into an automatic hamburger patty machine. Within a half an hour, he has 80 neatly rounded, ready-to-cook, four-ounce patties – more than enough when added to the two dozen uncooked patties from the day before.

Only a half hour before opening. Remembering the chili, he wipes the meat particles from the patty machine with the cloth, takes the large stock pot of chili from the refrigerator, fills a smaller container with enough for the lunch hour, and places the chili with the other items on the heated steam table.

Later, after the lunch rush is over, he allows himself a brief break. Sipping a cup of coffee, the hardworking owner is pleased, very pleased indeed, with his busy and profitable opening – not realizing he has exposed some of his patrons to possible food poisoning.

He should have been aware that raw chicken – like any meat – is a common carrier of Salmonella bacteria, a frequent cause of food poisoning. Proper cooking killed the bacteria on the chicken, but the owner didn't clean and sanitize the cutting board after portioning the chickens, nor did he wash his hands before handling the salads and hamburger. He compounded the problem by wiping the cutting board and patty machine with the same cloth, and then failed to dismantle, clean and sanitize the machine. He may even have

contaminated the cooked chicken with his hands after removing it from the oven. He also erred in putting the hot chili prepared the night before into the refrigerator in such a large stock pot. It should have been separated into smaller containers so that it cooled rapidly overnight. Further, half an hour was not enough time to reheat the chili to a high, and safe, enough temperature on a steam table. It should have been heated on a stove first.

This hypothetical restaurant owner made some common mistakes in the handling and preparation of food, placing both his customers and business reputation in jeopardy. The safe practices he should have followed are only a small part of what any skilled manager in a retail food establishment needs to know about proper food protection. To help ensure that managers have such knowledge is one of the main purposes of a new national effort called the Food Protection Certification Program – spearheaded by FDA – that will get under way in July. The program is designed to test and certify the thousands of men and women who supervise the hundreds of thousands of retail food operations that compete for the billions of food dollars spent by Americans away from their homes each year.

Such supervision places many demands upon a manager. For example, a manager should understand that causes and prevention of foodborne diseases, which in turn means knowing something about how the mishandling of various foods can lead to food poisoning. Food protection is an all-embracing phrase that also includes knowing what to do from the time foods are delivered to storing them properly, or shipping them to another destination, without spoilage. Training employees on their personal hygiene, dress, and direct handling of foods and being aware of any potential diseases an employee might transmit on the job is also a manager's concern. Good management demands knowing something about proper water and waste disposal, plumbing, cleaning and sanitizing procedures, maintenance and operation of equipment, and a sensitivity to the use of toxic materials and their separation from foods served to the public. And all of that is by no means an all-inclusive summary of what is meant by food protection and sanitation.

Arthur L. Banks, director of FDA's retail food protection branch in Washington, D.C., says the new Food Protection Certification Program is a joint effort of the "best minds" in the country on food protection and sanitation. The first objective in establishing the program was to create an examination to test all candidates on what they should know about proper food management and protection, one that would be acceptable to state and local regulatory agencies around the country, and one that was easy to understand, since the test would be taken by a variety of food management professionals – some with college degrees in their fields, others who participated in company or other training programs, and still others whose training was on-the-job.

Those passing the test will be certified and have their names placed on a national registry that backers of the program hope will be nationally recognized and accepted by the food industry and by state and local regulatory agencies (mainly health and agricultural departments) that are responsible for enforcing local health and sanitation codes. The registry will be updated regularly. A manager who fails the test will be given an opportunity for additional study or training.

The test – composed of 60 multiple-choice questions – was developed by a 40-member committee made up of food industry experts, educators, and federal, state and local regulatory officials. The committee, which was divided into five working groups, developed a total of 1,200 questions on food protection practices. Then a careful screening and evaluation process resulted in a test these experts felt could be used nationwide. A 17-member advisory and policy board, also composed of a cross section of public and private industry representatives, approved the test. The board functions as a permanent overseer of the program.

The program will be financed through fees paid by food sanitation training programs or by retail food managers or their employers. The Center for Occupational and Professional Assessment of the Educational Testing Service (ETS) at Princeton, N.J., conducts the tests. ETS is the largest independent, nonprofit testing organization in the United States. (The testing service's involvement in administering Scholastic Aptitude Tests – SATs – and financial-aid applications is well-known to college students and their parents.) ETS agreed to undertake and administer the Food Protection Certification Program, including maintaining the registry, after a survey last year indicated the program would have widespread support.

The test will be given at several thousand testing centers available to ETS, and individual managers taking the test generally will pay a $25 fee. If the test is part of a training program, the fee will be part of the overall cost of the program.

Although state and local regulatory agencies are responsible for enforcing health and sanitation codes in retail food establishments, FDA's Banks said that inspections alone, no matter how frequent, cannot ensure the safe and sanitary handling of food. "Government at any level will never be able to assure adequate food protection without the cooperation and day-to-day diligence of industry," he stressed.

"There are outbreaks of foodborne illnesses every year," Banks said. "People become ill and sometimes people die. We realize that passing a test will not assure that no one will get ill from the mishandling of food. We realize it takes more than knowledge to achieve conformance to food sanitation standards. But if an individual does not know what is expected of him or her or why, you cannot expect conformance and not have outbreaks."

A 1981 report by the U.S. Centers for Disease Control in Atlanta is indicative of the concern public health officials share over foodborne illnesses. The report showed that more than 14,000 persons were affected by foodborne illnesses in 568 outbreaks reported to CDC in 1981 (the most recent year for which figures have been compiled). Of that total, 251 of the outbreaks (44 percent) resulted from foods eaten in restaurants, cafeterias or delicatessens. (An outbreak is defined by CDC as any incident in which two or more people become ill after consuming a common food and where there is evidence that the food eaten was the source of the illness.)

CDC obtains its data from federal, state and local public health agencies, physicians and other sources, but the agency's report makes it clear that many more outbreaks occur than are reported. "The number of outbreaks of foodborne disease reported by this [CDC's] surveillance system represents only a small fraction of the total number that occur," the report noted. Also, outbreaks where illness occurs soon after eating are more likely to be reported than those where adverse effects are not felt for many hours, or sometimes even days or weeks later, because of the longer incubation periods of some illness-causing bacteria.

Food not heated or cooled to proper temperatures, mishandling of certain foods such as raw beef and other meat and poultry products, the use of contaminated equipment, and the use of foods from unsafe sources (such as shellfish from polluted waters) account for a large percentage of the outbreaks. The new testing program will have its main impact on the food service and retail food store industries. Together they comprise some 750,000 establishments and employ

10 million people. Eighty percent of those are employees in businesses and institutions that serve food. The others are in supermarkets, grocery stores, meat, fish and poultry markets, convenience stores, and other such businesses.

In general, says Banks, there is "as high a level of sanitation in our retail and food service establishments in this country as anywhere in the world." Even the equipment used has improved dramatically so that problems that once plagued health officials and operators years ago are rarely seen today. "For example," he said, "equipment like slicers, saws, and grinders that you could not clean because of the way they were designed and constructed then – that's no longer the case. On the other hand, there is always a percentage of operations which are poorly managed and which pose potential health risks. FDA supports the view that individuals who manage food operations should understand the basic concepts of food protection and sanitation."

A logical question, then, is where do operators and managers of these businesses learn these concepts? There is today a wide network of training and testing programs. Many food chains and other large companies have their own programs. Others are sponsored by industry and trade groups, or by state and local health and agricultural departments. Universities and community colleges often offer food sanitation and protection courses, and Banks predicted that many educational institutions would adopt the new test as part of their course offerings.

In the view of many regulatory officials and others in industry, Banks said, many training and testing programs are often "poorly focused." That is, some emphasize the wrong things, they're not always consistent in what they teach, and they don't always measure what a person should know or doesn't know, he said. Only a few state and local jurisdictions make participation in a training and testing program mandatory, says Banks. "There are some 3,000 state and local regulatory agencies in the United States, and in 99 out of every 100 local jurisdictions, it is voluntary," he said.

Although FDA has no plans to make the new testing and certification program mandatory – leaving that decision to state and local regulatory agencies – there has been industry opposition, as well as support, for the program. The National Restaurant Association objected last year to the certification of managers, saying the program "would burden food service operators with increased costs and cause unneeded government regulation without increasing the current level of food protection and sanitation to the public." Failure "to pass the test...could also deprive food service operators of their livelihood," the association stated.

FDA's Banks said that industry's cost of training and certifying a manager would not change significantly as a result of the program. Many large companies already make substantial investments in training courses for managers and materials for employees, others pay for training courses at the community level. In addition, local agencies incur costs in monitoring existing industry training and testing programs. Under the new program, ETS will provide the test materials, score the tests, inform candidates who do not pass why they failed, maintain and update the national registry and other records, and handle other administrative costs through the fees collected – an attractive feature to many budget-conscious local jurisdictions.

There is already widespread interest in the new testing and certification program. "We know of more than 300 training programs in existence today, and more than 100 sponsors already have told FDA that they will use the [new] program as soon as the examination is ready to be used," Banks said.

"When this test becomes available," he continued, " it will be required by more people, not only because it will be company policy of many firms to have employees take the test and be certified, but it will be required by some regulatory agencies that haven't previously done so." However, Banks said that FDA was "neutral" on whether local jurisdictions should make participation in the program mandatory. He predicted that the largest increase in the use of the new program would be among those who voluntarily elect to use or take the test.

Some industry sources question whether it is reasonable to expect that any single test can be useful for both food management professionals with college level training and others with no formal training. "We have had companies call and say they have the best managers in the country, people who completed a rigorous training program that the company intended to be tough, and now you're asking them to pass only a test that any Mom-and-Pop operation on the street must also pass," Banks said. "On the other hand, we have people who argue this test could put small operators out of business because they'd have to be college grads to pass it.

"Well, it doesn't matter whether you're a university graduate or a Mom-or-Pop type. If you don't have the knowledge, you aren't going to pass the test. It's not something you can pass if you don't have the information. On the other hand, there is no reason why you can't assess whether that person has the needed information without using college level jargon. You can ask the questions in plain, simple English, and if the Mom-and-Pop types have the knowledge and can read the question, they should be able to successfully take the test." ◆

Chris Lecos is a member of FDA's communications staff.

Food Inspection Results

NEW YORK TIMES, NOVEMBER 29, 1987, P.71.

The New York City Department of Health last week released a list of 24 food establishments cited for violating the health code.

In addition, it listed eight restaurants that were allowed to reopen after correcting earlier violations, and four that were closed for having failed to do so.

The establishments are these:

Violations
• MANHATTAN

Atithi Indian Restaurant, 67 University Place (10th St.). Operating without a valid permit.

Bubba's, 17 University Place (Washington Mews). Food on floor, mouse droppings on kitchen work tables, inadequate extermination, encrusted can opener.

Dallas Fried Chicken, 104 Lenox Ave. (115th St.). No food-protection certificate, broken tiles, dirty ceiling tiles, shelves made of porous materials.

Hisae's Restaurant, 35 Cooper Square (5th St.). Mouse droppings on kitchen work table, flies in basement, inadequate extermination, inadequate sanitization of flatware and utensils, no food-protection certificate.

La Croissant, 40 W. 8th St. (MacDougal St.). Operating without a valid permit, no food-protection certificate.

Snack Bar Express, Passageway Grand Central Terminal. Hot food kept too cold.

• BROOKLYN

Anthony's Luncheonette, 163 Montrose Ave. (Graham Ave.). Rat droppings on rear floor and at floor-wall junctions in cellar.

Casbah, 1296 Nostrand Ave. (Clarkson Ave.). No food-protection certificate.

Deli-Restaurant, 715 Flushing Ave. (Throop Ave.-Thornton St.). Operating without a valid permit, hot water supply not provided, materials stored directly on floor, rat droppings behind service counters.

Granada Restaurant, 1126 Nostrand Ave. (Maple St.). Operating without a valid permit, no food-protection certificate, extermination report unavailable, greasy ceilings and walls near exhaust fan, flies in kitchen.

International Restaurant, 131 Tompkins Ave. (Vernon Ave.). Operating without a valid permit, no food-protection certificate, no immersion basket and gas burner under sink system for sanitization, last inspection report unavailable, hot food kept too cold, toilet broken.

Le Banquet Restaurant, 3405 Church Ave. (E. 34th St.). Operating without a valid permit, chipping wooden floors in kitchen prevent proper cleaning, final rinse water too cool, hot foods kept too cool, kitchen ceiling broken over box refrigerator, holes in kitchen floor and walls, greasy kitchen floor-wall junction, hot holding table used to raise food temperatures, roaches in kitchen.

Mirage Restaurant, 2670 Coney Island Ave. (Ave. X). Dishwasher broken and sanitization facilities unavailable, food trays in walk-in refrigerator, raw meat stored on top of cooked food, food handlers not wearing hairnets, inadequate extermination, open kitchen cutting boards, encrusted interior walls of refrigerator, dirty and encrusted shelves in walk-in box, last inspection report unavailable, no self-inspection records kept.

Our Place, 492 Broadway (Rutledge Ave.). Operating without a valid permit, rat droppings in food preparation and rear storage areas, kitchen floor-wall junction greasy and dirty.

Rey's Restaurant, 949 Grand St. (Catherine St.). Flies in kitchen and service areas, choking prevention poster not posted, hand wash sign not posted in employee bathroom, floors in kitchen and service areas inappropriately covered, soap and paper towels unavailable in employee bathroom.

• BRONX

Danubio Restaurant, 109 W. 225 St. (Bronx River). Flies in food preparation area and cellar, inadequate extermination, encrusted floor in food preparation area, dishes and flatware not sanitized, no food-protection certificate, stored containers used to serve customers.

• QUEENS

Yeh's Wok, 99-13 Queens Blvd. Mouse droppings on kitchen and floor storage area, alcohol-birth defects sign not posted, employee handwash sign not posted in bathroom, no screens in kitchen and second-floor food preparation area, food, stock and supplies stored directly on floor.

La Colombiana, 81-12 Roosevelt Ave. Employee hand-wash sign not posted in bathroom, dirty cellar floor, dirty and dusty floor-wall junctions, final rinse temperature too cool, sink blocked, sink overflows onto the floor, mouse droppings on cellar storage bin shelves, flies in the cellar, operating without a valid permit.

Burger King, 162-31 Jamaica Ave. Flies in the service area.

Chu Chu's, 92-15 Merrick Blvd. Some cold food kept too warm, some hot food kept too cold, no thermometer at service counter refrigerator.

Chow's Dynasty, 84-22 37th Ave. Paint peeling on walls and ceiling, no immersion basket for sanitization, and gas burner is under second sink compartment instead of third.

Nino's Pizzeria, 179-19 Hillside Ave. Mouse droppings on basement shelves and at floor-wall junctions, kitchen refrigerator encrusted, food uncovered in kitchen reach-in refrigerator, refrigerator fan dusty, greasy kitchen can opener.

• STATEN ISLAND

Cardini's, 142 Donagan St. Final rinse too cool, no device to prevent vacuum in hose, operating without a valid permit.

Kentucky Fried Chicken, 2189 Hylan Blvd. Flies, mouse droppings on floor under kitchen sink, extermination report unavailable, portions of kitchen wall unpainted, wall broken and unpainted at baseboard, ground area at outside garbage, meter greasy, dirty sewer pit in storeroom, broken screen at rear door, peeling paint on walls and ceiling of walk-in box, rusty area on top of ice machine, operating without a valid permit.

Reopened
• MANHATTAN

Bright Fortune Villa, 270 W. 38th St.
Cathedral Cafe, 941 Amsterdam Ave.
El Quijote, 226 W. 23rd St.
Louis Restaurant, 441 Amsterdam Ave.

• BRONX

Cafe Cappucino, 1036 Morris Park Ave.

• BROOKLYN

Circle Cafe, 514 2nd Ave.

• QUEENS

Gazaly Deli, 601-03 Beach 20th St.
Lotto Luncheonette, 96-22 Queens Blvd.

Closed
• MANHATTAN

Little Italy, 208 W. 42nd St.
Pizza Number One, 1741 Broadway.

• BROOKLYN

El Nuevo Dugout, 168 Marcy Ave.

• QUEENS

M & M Luncheonette, 119-07 Liberty Ave.

[NOTE: Data as of November 29, 1987.]

Some Chemicals Found In Food

AMERICANS FOR SAFE FOOD
CENTER FOR SCIENCE IN THE PUBLIC INTEREST

Captan – a fungicide that causes cancer in lab animals... used on various fruits, nuts, and vegetables including apples, peaches, almonds, strawberries, beans, peas, beets, carrots, corn, garlic, cabbage, lettuce, potatoes, kale, spinach, and broccoli... has been found in oils, fats, and shortenings... used in post-harvest treatments for potatoes and fruit, sometimes in packing boxes... used as a seed treatment on corn and soybeans... toxic to fish, bees, earthworms, and birds.

Parathion (ethyl parathion) – an insecticide that causes birth defects in chicks and ducklings... evidence suggests it could be carcinogenic in laboratory animals... used on cotton, tobacco, and many fruit and vegetable crops... banned in six other countries... direct exposure extremely hazardous... highly toxic to birds, bees, and other non-pest species.

Daminozide – trade name Alar, manufactured by Uniroyal Co... applied as a plant-growth regulator to apple and other orchard crops (cherry, nectarine, peach, prune, pears, grapes) to increase the fruits' storage life, firmness, and coloring... also approved for use on peanut crops... systemic chemical penetrates the whole fruit and cannot be removed by washing or peeling... causes cancer in laboratory animals... a breakdown product (unsymmetrical dimethylhydrazine, UDMH) is an even more potent carcinogen.

Paraquat – a potent herbicide used to prepare fields for planting or harvesting... used on soybean crops as a harvest aid... extremely poisonous on direct exposure... toxic to fish and other species.

Benomyl – a post-harvest fungicide used on apples, peaches, cherries, plums, apricots, nectarines, pears, pineapples, bananas, mangoes, etc... linked to malignant tumors and birth defects in laboratory animals... highly toxic to earthworms and some species of fish.

Methyl Bromide – used as a pre-plant soil fumigant and as a post-harvest fumigant to control insects on stored nuts, grains, fruits, and vegetables... very toxic on direct exposure – 60 deaths since 1955... Environmental Protection Agency (EPA), awaiting results of studies regarding cancer, birth defects and mutations, nevertheless recently allowed use of methyl bromide on macadamia nuts, fava beans, sweet potatoes, pistachio nuts, kiwi fruit, cereals, flour, spices, root crops, lentils, and leafy vegetables.

Trifluralin – also known as Treflan... contains a suspected carcinogen, NDPA... commonly used on carrots, soybeans, cotton, wheat, and barley... EPA awaiting more data on health risks.

Chlorothalonil – trade name Bravo... a fungicide found to cause tumors in rats and mice... used on various vegetables, fruits, and beans including tomatoes, onions, broccoli, cabbage, cantaloupe, carrots, cauliflower, celery, cucumber, lettuce, potatoes, and watermelon... highly toxic to aquatic invertebrates.

Linuron – an herbicide used on crops including soybeans, carrots, celery, asparagus, corn, potatoes, and wheat... associated with tumors in rats and mice... contaminates ground and surface water supplies.

Aflatoxin – a powerful carcinogen produced naturally by a mold that grows on crops including field corn and peanuts... human exposure primarily from peanuts and milk from dairy cows raised on contaminated feed... one experiment produced tumors in all rats eating food contaminated with 15 parts aflatoxin per billion.

Clorsulon – used to treat livestock afflicted with a type of parasitic flatworm called a fluke... a suspected carcinogen (kidney tumors).

Fenbendazole – used to treat cattle for parasites... a mutagen and a possible carcinogen.

Thiabendazole (TBZ) – used to treat livestock for parasitic intestinal worms... a suspected carcinogen... also used for post-harvest treatments of bananas, citrus, apples, pears, sugar beets, and potatoes to protect against mold during storage... appears to cause fetal abnormalities and fetal death in laboratory animals.

Antibiotics – added to feed of beef cattle, dairy cows, swine and poultry to increase growth rates and prevent disease... use leads to drug-resistant "supergerms" that are harder to kill when they cause infections in humans... cases of food poisoning caused by drug-resistant germs are on the rise... less crowding in feedlots would lessen the need for antibiotic use.

Gentian Violet – a carcinogen... still used to treat chickens and turkeys for diseases caused by fungus... added to poultry feed to inhibit mold.

Nitrofurans (Nitrofurazone and Furazolidone) – carcinogens... used to treat livestock disease, but also added to feed in order to increase growth rates, particularly for swine and poultry.

Aldicarb – trade name Temik, manufactured by Union Carbide... a pesticide used on beans, cotton, citrus fruit, peanuts, pecans, potatoes, sorghum, soybeans, sugar beets, sugar cane, sweet potatoes, bananas, coffee, and others... a frequent contaminant of drinking water... disrupts nervous system at high doses... potatoes are a main source of exposure... summer of 1985: California officials order the destruction of millions of watermelons after the illnesses of 180 people were traced to aldicarb residues in melons, which appeared normal and had no unusual taste or smell... found to substantially weaken the immune systems of mice exposed to extremely low levels of aldicarb in their drinking water... altered the immune function of women who drank groundwater contaminated with minute amounts of aldicarb.

America's Pesticide Permeated Food

BY ANNE MONTGOMERY
NUTRITION ACTION HEALTHLETTER, JUNE 1987, PAGE 1

Conventional industry and government propaganda holds that America has the "safest food supply in the world." And most supermarkets, with their aisles of colorful processed foods, overflowing freezers, and meticulously arranged fresh produce, reinforce this notion in millions of consumers every day.

But a lot of what's sold isn't safe at all. Pesticide-contaminated food has become a topic of growing concern, as overuse, misuse, and abuse of these toxic chemicals has escalated. According to a recent survey conducted by the Food Marketing Institute, "Three out of four consumers consider residues such as pesticides and herbicides to be a serious health hazard." Unfortunately, the very foods that health-conscious consumers are trying to eat more of – fresh fruits and vegetables – are those that are most likely to be contaminated.

On July 4, 1985, millions of people scattered throughout California and Oregon laughed, joked, and watched firework displays. They brought picnics – snacks, sandwiches, cold drinks, salads, and fruit for dessert. Within hours, hundreds of those who consumed watermelon became violently ill, developing symptoms ranging from nausea and diarrhea to seizures, blurred vision, and irregular heartbeat. Belatedly, the Food and Drug Administration (FDA) discovered that a large number of California watermelons were contaminated with the acutely toxic insecticide, aldicarb.

Reports of illness associated with eating watermelons continued to pour in through most of July, bringing the total to 1,350. When laboratory analyses were run on the contaminated melons, it was found that, in many cases, residue levels were well *below* those that FDA's routine screening tests can detect.

For officials at both FDA and the Environmental Protection Agency (EPA), the most worrisome aspect of the contamination outbreak was the question of how aldicarb ever got into watermelons in the first place. An insecticide primarily used to treat citrus and potato crops, aldicarb cannot legally be used on melons. One theory holds that the melons absorbed the chemical from tainted groundwater, which has been known to remain contaminated with the pesticide for years. The more plausible explanation was illegal use by growers.

In another well-publicized incident, unacceptably high levels of daminozide, a plant growth regulator, were found in apples grown in the Northwest. Even more worrisome, a highly carcinogenic breakdown product of daminozide, UDMH, was found at high levels in processed apple products, such as applesauce (some forms of processing concentrate pesticide residues, or their breakdown products).

Reaction to the contamination was swift. Beginning with Safeway, last summer several major supermarket chains told apple growers they would not accept further shipments of daminozide-treated apples. Manufacturers of processed apple products, including Gerber, Heinz, Grand Union, and Welch also boycotted treated apples.

Protecting the public?

Two agencies, the EPA and the FDA, share joint responsibility for protecting consumers from the effects of hazardous pesticides. Yet, as the above examples illustrate, they offer little more than Swiss-cheese protection. The most vulnerable of all, farmworkers and their children, are virtually at the mercy of these poisons. Not only are farmworkers exposed – like the rest of us – to residues in food, but they also handle pesticides directly, and inhale the chemical as it is sprayed on crops. Farmworkers also tend to live near fields. This increases the likelihood that their drinking water (particularly if it comes from a well) will be contaminated with pesticides.

Richard Wiles, an agricultural policy analyst at the National Academy of Sciences (NAS), thinks that the health of the country's 472,000 farmworkers is seriously jeopardized by pesticides. EPA "pretends that everybody, even field hands who work in an atmosphere permeated with pesticide residues, absorbs no more than the minute quantities allowed in food," he charges.[1]

A growing body of medical research points out the hollowness of this assumption. A 1986 study of Kansas farmers, for example, shows that the relative risk of non-Hodgkin's lymphoma (NHL), a cancer of the lymphatic system, increased significantly with the number of days of herbicide exposure per year. Men exposed to herbicides more than 20 days per year had a risk of NHL six times higher than non-farmers.[2] Frequent users who mixed or applied the herbicides themselves had an eight-fold increase.

In a California study, children born to agricultural workers were found to have deformed limbs at a rate of 5.2 per thousand – 13 times the rate among the general population. [3] Unfortunately, it's probable that these findings are just the tip of the iceberg. "No one," notes Wiles, "is quite sure how many people in the United States are poisoned by pesticides each year, since most field hands don't seek medical treatment unless they become incapacitated." [1]

Fraudulent data

Perhaps the greatest obstacle the government has faced in regulating pesticides is that it has had to work with inadequate or fraudulent health and safety data – data that were submitted to support the registrations for many pesticides. Before a pesticide can be sold, it must first be registered with EPA under the Federal Insecticide, Fungicide, and Rodenticide Act (FIFRA).

Adopted in 1947 and rewritten in 1972, FIFRA charges EPA with insuring that pesticides "will not generally cause unreasonable adverse effects on the environment," including human beings. Agency scientists review health and safety data supplied by manufacturers and set a "tolerance" level, or a maximum allowable limit, for food residues. They also decide how much of each chemical should be applied to a particular crop. FDA enforces the tolerance levels through its residue monitoring program.

But in 1976, it became clear that the tolerance-setting system was severely flawed, because it had presumed that manufacturers had submitted accurate and scientifically valid health and safety data. FDA investigators accidentally discovered that one of the largest testing laboratories in the world, Industrial Bio-Test Laboratories (IBT) of Northbrook, Illinois, had for years been producing fraudulent data for industry. Ultimately, a large part of IBT's data was discredited. Yet EPA allowed the registrations of over 200 pesticides – 90 of which are used to treat food crops – to remain in place, even though they were based to some degree on the scientifically discredited IBT data. [4]

Consumers fight back

The history of poor testing and sloppy regulation for pesticides is finally beginning to arouse the public's ire. Groups such as the National Coalition Against the Misuse of Pesticides (NCAMP) and the Natural Resources Defense Council (NRDC) are making their opinions heard. As NRDC's Karen Snyder says, "Pesticides affect us all. Although farmworkers are most vulnerable to their effects, all of us are exposed indirectly – and involuntarily – through the food we eat. Not all pesticides are unsafe. But some can cause cancer, birth defects, heritable genetic mutations, and nerve damage. Clearly, we need to spend more money and devote more resources to improving our testing procedures for detecting hazardous residues."

Contributing to the pesticide problem is the fact that neither processed foods nor bulk fresh produce are labeled with pesticide "ingredients." "The consumer has no way of knowing which pesticides were used during growing and processing, much less whether they might pose health hazards if ingested at low levels over an extended period of time," says Deborah Schechter, director of CSPI's Americans for Safe Food project. "If consumers did know, they could influence what's used by not buying produces treated with certain chemicals."

Granted, it would be difficult for industry to list every single pesticide used on produce, since at the wholesale level, variously treated batches from around the country (or the world) are mixed together. But in one case, providing consumers with pesticide information would be feasible. Every day, when produce is unloaded off trucks and arranged in supermarket displays, workers discard the crates labeled with the names of those fungicides and fumigants that were added after the fruit was harvested. While postharvest pesticides are required by the Federal Food, Drug and Cosmetic Act (FDCA) to be listed on shipping crates, the law does not require that consumers must also see this information.

Another disquieting fact is that scientists know surprisingly little about the effects of low-level exposure to pesticides, and therefore doctors are rarely able to diagnose pesticide-related illness. Nor do scientists understand the effects of breakdown substances the body creates when it processes, or metabolizes, pesticides. So-called "inert" ingredients in pesticide products, such as solvents, colorings, and preservatives, are almost never tested for toxicity – and are exempt from residue tolerance requirements. Finally, no one knows what chemical interactions occur between different pesticides when several are used on the same crop, or what happens when a combination of pesticides is ingested by the same person.

Please don't eat the cantaloupe

As if those "unknowns" weren't enough to make fresh fruit and vegetable lovers shudder as they bite into their next apple, FDA bases its tolerance levels for many pesticides on wildly inaccurate estimates of how much produce some Americans eat. Take cantaloupe, for example. FDA's "average" diet assumes that Americans eat no more than 7.5 ounces - about one-half of a single melon – per year. Similarly, the agency assumes we eat only one avocado, 1 1/2 cups of cooked summer squash, 2 1/2 tangerines, one mango, 1 3/4 cups of cooked winter squash, and 1 1/2 cups of cooked brussels sprouts, among dozens of examples.

In other words, eating more than 7.5 ounces of these fruits and vegetables could conceivably expose the consumer to a level of risk above what the government considers "safe."

The government regulates pesticide residues in processed foods somewhat differently than it does the residues in fresh produce. EPA sets tolerance levels for pesticide residues in those processed foods, such as tomato paste and tomato sauce, that contain a higher concentration of pesticide than their fresh counterparts. FDA treats residues in these processed foods as additives; not so in fresh. Only food additives are subject to the strictures of Section 409 of the FDCA, which includes the famous Delaney amendment: "No additive shall be deemed to be safe if it is found to induce cancer when ingested by man or animal."

As updated toxicity data on "older" pesticides (those that have been routinely used for years) comes in, it is showing that many of the pesticides that are present in processed foods are cancer-causing. Since the Delaney clause specifically outlaws carcinogenic additives, EPA must – theoretically – cancel the registrations of all pesticides with potentials to induce cancer, if they occur in processed foods. But so far, the agency has backed off from carrying out this responsibility, claiming that "when the pesticide meets the FIFRA standard, but not the Delaney standard, the proper course of action is obscure."[5]

Missing evidence

While pesticides that concentrate in processed foods are considered food additives, residues on raw agricultural commodities – primarily fresh produce and grains – are not. Therefore, cancer-promoting pesticides can be, and are, used to treat these foods. Ironically, the data to support the carcinogenic potential of certain pesticides comes from the industry itself.

In the mid-1970s, Congress chided EPA for having too little health and safety information on many widely used pesticides. After carefully evaluating its files, the agency discovered it lacked adequate long-term health effects data on approximately 600 active ingredients (out of a total of about 1,200).[6] EPA informed pesticide manufacturers that their registrations would be revoked unless safety data, meeting modern premarket testing requirements, were submitted to fill data gaps or replace invalid data. Faced with the option of complying with EPA or going out of business, the industry reluctantly went to work.

But according to a 1984 NAS report entitled, "Toxicity Testing: Strategies to Determine Needs and Priorities," 71 to 80 percent of all pesticides sold in the U.S. had not been sufficiently tested for cancer. For genetic mutations, 21 to 30 percent of all pesticides had not been fully tested; for birth defects, the figure

was 51 to 60 percent; and for adverse effects on the nervous system, 90 percent.

The bottom line

For the consumer, the big question is, how safe is the food supply now? The answer is, unfortunately, not safe enough. FDA has a monitoring program that supposedly enforces EPA's residue tolerance levels by screening samples of domestic and imported produce. But FDA's program has been sharply criticized in numerous reports issued by the General Accounting Office (GAO), the investigative arm of Congress.

One recently released report criticized the way FDA tests imported foods. It pointed out that the agency's screening tests are capable of detecting only 203 out of roughly 400 pesticides. The report also noted that FDA knows little "about actual pesticide use in foreign food production.... In addition to a lack of knowledge about pesticide chemicals manufactured and used overseas, FDA lacks country-by-country information about the ultimate destination and use of pesticides manufactured in the United States and exported to other countries." At an April 30 Congressional hearing on the problem, for example, Congressmen were distressed to learn FDA has not been testing imported produce for EBDCs, a widely used family of post-harvest fungicides.

Furthermore, said the GAO, because food might spoil while awaiting drawn-out testing, FDA rarely holds up sales of imported produce. As a result, by the time a sample of produce has been identified as contaminated, the shipment from which it came has already been sold – and presumably eaten. GAO estimates about 6 percent of the imported produce that's annually sampled breaks the law because it contains pesticide residues above the tolerance level.

Another recent GAO report examined FDA's program for monitoring residues on domestic produce. These conclusions, too, are far from reassuring. According to GAO estimates, "FDA samples less than *two-tenths of one percent* of domestic food production [emphasis added].[8] Moreover, the report noted, "FDA is not regularly testing for a number of pesticides" that, by the FDA's own admission, "require continuous or periodic testing because of their potential health hazard and likely usage." For domestic produce, GAO estimates the violation rate is 3 percent.

There are no quick solutions to the pesticide dilemma. But given the pervasiveness of these chemicals in both our food supply and the environment, it is clear that more testing and research are needed – which translates into making pesticide surveillance a higher priority in the budgets of both FDA and EPA. Apparently, senior managers at EPA agree: A recent agency report entitled, "Unfinished Business: A Comparative Assessment of Environmental Problems," ranked pesticide residues in food as the number three cancer risk, after workplace toxins and radon in the home. "Risks and EPA's current priorities do not always match," the paper noted.

How about labeling? Addressing this question, Bill Schultz, of the Public Citizen Litigation Group, says, "A good argument can be made that the law *already* requires labeling in processed foods where residues are treated as food additives."

For fresh produce and grains, which are not covered by labeling laws or the Delaney amendment, FDA should monitor residues more closely and frequently to identify those foods that are contaminated with hazardous pesticides. The cost of developing a fast, practical lab test to detect a particular pesticide in foods could be borne by that chemical's manufacturer.

One thing is certain: Controversy over pesticide residues – on both imported and domestic foods – is intensifying. According to David Pimentel, Professor of Insect Ecology and Agricultural Sciences at Cornell University, in the last 40 years, the volume of pesticides used in American agriculture has increased tenfold. At the same time, it's becoming clear to more and more farmers that pesticides are not panaceas. Despite the huge increase in insecticide use since World War II, for example, crop losses from insect pests have increased from 7 percent to 13 percent.[9]

It's time to turn back the clock and take a close look at safer, non-chemical methods of growing, storing, and preserving crops – *before* the environment and the food supply become even more contaminated with persistent, toxic residues. "There's a growing feeling that organic farming is the most promising permanent solution to the pesticide problem," comments CSPI's Schechter. "State departments of agriculture and agricultural universities should give it top priority."

Pesticide residues, animal antibiotics, and microbial food contaminants are covered in *Guess What's Coming to Dinner,* a new 50-page booklet written by the national staff of Americans for Safe Food.

References:

1. Richard Wiles, "Pesticide Risk To Farm Workers," The Nation, October 5, 1985.
2. J. Am. Med. Assoc. 256: 1141, 1986.
3. Schwartz et al, Physician's Handbook on Pesticides, San Diego: California Public Interest Research Group and the Department of Community Medicine, University of California, San Diego Medical Center, 1980.
4. Lawrie Mott with the assistance of Martha Broad, Pesticides in Food (Natural Resources Defence Council, Inc.) 1985.
5. Testimony of EPA Assistant Administrator John A. Moore, April 7, 1987, Subcommittee on Department Operations, Research, and Foreign Agriculture of the House Committee on Agriculture.
6. Testimony of Lawrie Mott, April 30, 1987, Subcommittee on Oversight and Investigations of the House Committee on Energy and Commerce.
7. General Accounting Office, Pesticides: Better Sampling and Enforcement Needed on Imported Food, 1986.
8. General Accounting Office, Pesticides: Need to Enhance FDA's Ability to Protect the Public From Illegal Residues, 1986.
9. Bioscience 28: 772, 1978.

PCBs and Breast Milk –
Mothers Reconsider Choice of Nursing Over Bottled Formulas Given PCB Threat
BY MARCELLA R. MOSHER AND GREG MOYER
NUTRITION ACTION HEALTHLETTER, NOVEMBER, 1980

Breast milk, our species' nutritional link with the newborn, is polluted.

That news came as a shock back in 1976 when toxic industrial chemicals were found in virtually all samples of mothers' milk tested by the Environmental Protection Agency (EPA). Polychlorinated biphenyls – the notorious PCBs – had seeped into our food chain and in a scant half-century became an extraordinarily widespread chemical contaminant.

To environmentalists, nutritionists, and progressive health advocates this news was a devastating blow. Just as America was weaning itself off bottle-feeding and artificial supplements, the purity of the "ideal food" seemed seriously compromised.

The public learned that PCBs – shown to cause cancer in laboratory rats and mice – were widespread in the natural environment. Air, water, soil, plants, and animals were all touched. Coincidentally, a study from the University of Wisconsin Primate Research Center reported that offspring of monkeys nursed with PCB-laden milk exhibited skin disorders, showed signs of hyperactivity, and appeared to have learning deficiencies. Within a year, three of the six infant monkeys had died.

The public outcry then, as now, was one of anger mixed with frustration. No woman can quickly or totally cleanse her system of this toxic industrial product. Nor can she totally avoid ingesting some amount of the substance from the food she eats.

People slowly realized that future generations were destined to be chemically "branded" with one of the most toxic and persistent synthetic chemicals ever invented. The PCB tragedy became the classic example of what happens when toxic chemicals are let loose in the environment without strict governmental controls.

But the possible consequences of PCBs' effects are compounded for infants who breast-feed. Research has shown that breast-feeding tends to pass on higher concentrations of PCBs and other toxins than had existed in the mother. As a woman lactates, she draws on nutrients stored in her fatty tissue. And it is in these cells where the pollutant accumulates.

Once the PCBs get into the blood, they quickly find their way into the mother's milk. Ironically, one of the best methods for excreting the poisons is through breast milk.

The Wisconsin primate study showed that it took only three months for the breast-fed monkeys to reach the contamination level of their mothers. By the end of the lactation period, the levels in the babies had far exceeded the levels of the mothers.

Aware of this new scientific information, parents are put in the position of gambling with their infant's well-being. A mother can elect to breast-feed given the ample nutritional, sensory, and psychological benefits, but risk exposing her baby to higher levels of PCBs. Or she can opt for bottle-feeding and settle for the lower nutritional content of the supplements in order to guard the infant from the toxins.

How Did It Happen?

PCBs belong to a group of chemicals known as chlorinated hydrocarbons. This highly stable organic compound arrived on the scene in 1927 when Theodore Swann, of the Federal Phosphorus Company, first used biphenyl in the refining of lubricating oils. Two years later, Swann invented polychlorinated biphenyls and sold the process to the Monsanto Company.

Since 1930, 1.25 billion pounds of the stuff have been manufactured for use in this country. Over half of all PCBs ever made are still in use, though the 1976 Toxic Substances Control Act allowed EPA to order all further production halted. About 90 percent of the rest lies in dumps, landfills, soil, water, and living organisms. The EPA estimates that about 10 million pounds contaminate the biosystem each year through seepage, spills, and evaporation. For years PCBs were routinely discharged into rivers, streams, and sewers.

In its 50-year history, industry has found no shortage of uses for this versatile material. It burns only at very high temperatures, conducts heat well, refuses to dissolve in water, and resists corrosive chemicals. Only in fats will the PCB molecule enter solution. The compound has been used in everything from transformers and power capacitors, adhesives, washable wall coverings, upholstering materials, and flameproofers to early versions of carbonless copy paper.

PCBs are actually many times more toxic and much slower to break down into less harmful forms than the better known insecticide pollutants: DDT, chlordane, dieldrin, and mirex. The toxic effect of the biphenyls was first seen in 1943 among workers handling electrical equipment who had developed skin lesions and liver damage. In 1966, PCBs were found in birds in Sweden.

PCBs created more headlines in the early 1970s when striped bass caught in the Hudson River near New York City registered dangerously high concentrations. By 1974, the U.S. Department of Interior's Fish and Wildlife Service identified PCBs in 90.2 percent of freshwater fish sampled across the entire United States.

Each step up the food chain tends to concentrate the toxins. This process is known as "bioaccumulation." For example, fish have been discovered with PCB levels many million times that of their watery environment. Also, cow's milk contains

PCB Contamination in Breast Milk

SUBJECT	TOXINS IN MILK	CONTAMINATION BY WEIGHT*
Monkey	16 ppm PCBs on a fat basis (Toxicological effect observed at Wisconsin Primate Center)	64 micrograms of PCB per kilogram of body weight.
Infant (human)	1.8 ppm PCBs on a fat basis. (EPA national average)	10 micrograms of PCB per kilogram of body weight.
Infant per (human)	10.6 ppm PCBs on a fat basis. (Worst case uncovered by EPA study)	60 micrograms of PCBs kilogram of body weight

*These figures are based on the assumption that baby monkeys weighed 500 grams and drank 100 cubic centimeters (cc) milk per day, while human babies weighed six kilograms and drank 1,000 cc milk per day.

Data drawn from Harris, "Health Risks and Benefits of Nursing or Bottlefeeding: The Limits of Choice," a chapter in the forthcoming book, *Strategies for Public Health.*

a much higher PCB level than the corn on which the animal feeds. Breast-feeding, in effect, sends bio-accumulation into overdrive.

Who Are the Victims?

It falls to the tiniest and weakest members of the human race to withstand the greatest concentrations of PCBs. The studies of Dr. James Allen, formerly of the Wisconsin Primate Center, and new research detailing the transfer of PCBs through breast milk underscore the dangers for the newborn in a post-PCB world.

But another category of victims include those people exposed by accident or through occupation to unusually high dosages of PCBs. The real life experience of a Japanese community serves as a frightening example of the awesome power packed in this toxin.

In 1968, residents on the Japanese island of Kyushu unwittingly ate large amounts of PCBs when a cooling pipe leaked into a vat of rice oil at a food processing plant. The level of contamination (2000 to 3000 parts per million) reached 200 times what is considered safe by the U.S. Food and Drug Administration. Officially, 1,291 people contracted the "rice oil disease" characterized by acne-like rashes, headaches, nausea, diarrhea, loss of hair, loss of libido, fatigue, numbness, and menstrual disturbances.

Thirteen children were born to PCB-exposed women. One fetus was stillborn, four were small for gestational age, ten had dark skin pigmentation, four had pigmented gums, nine had conjunctivitis (inflammation of the membrane around the eye), and eight had neonatal jaundice. Nine years after the accident, a follow-up study of some of the children breast-fed by exposed mothers showed "slight but clinically important neurological and developmental impairment," according to a report of the research prepared by Dr. Walter J. Rogan of the National Institute of Environmental Health Sciences, and others.

Allen's rhesus monkeys fared little better. His research involved feeding controlled amounts of PCBs to female monkeys. Their offspring exhibited the same skin disorders as the victims of the "rice oil disease." The baby monkeys failed to make the developmental progress shown by the control group and several died at an early age.

A study led by Dr. Hirokadzu Kodama of Nagoya, Japan, finds comparable effects in humans. He discovered that mothers with average exposure to PCBs are still likely to pass on levels higher than their own to nursing offspring thanks to bio-accumulation.

Still, some researchers caution against making direct comparisons between monkeys and humans, because the monkeys have less subcutaneous fat in which to store the PCBs. In the animals, the toxins assault the vital organs more quickly. But with this caveat in mind, one can compare the levels of PCBs ingested by a nursing infant and those which were shown to affect the health of the monkeys.

The average exposure infants receive when adjusted for bodyweight is one-sixth that shown to adversely affect the baby monkey's health. In the worst case found by EPA's nationwide sampling, a human infant was receiving virtually the same dose as the monkey. Normally, regulatory agencies use a safety factor of 100 to set a "safe" standard for humans based on the results of the animal studies. Here, the concentrations are on the same order of magnitude, if not actually equivalent when translated to the common denominator of micrograms per kilogram of body weight.

The EPA figures are shocking in another sense: the average levels of PCBs in human breast milk is seven times higher than the amount permitted in cow's milk by the Food and Drug Administration. And it is about ten times what the FDA recommends as an acceptable daily allowance for infants.

In a telephone interview, Allen underscored the need for parents and the pediatrician to know about the physiological changes that may be occurring in infants exposed to high levels of PCBs. He said autopsies of the monkeys indicated abnormalities of the lymph system, bone marrow, thymus, and cortex. In short, he saw irregularities in some of the organs most important to normal growth and development.

Despite these ominous signs, Allen does not suggest women return to bottle-feeding their babies: "The values of breast-feeding outweigh the deleterious effects," he said. "I would advise my wife to breast-feed unless we lived in an area of high exposure to PCBs."

Regulators Mum on Risks

In its report to the nation, the Surgeon General's *Healthy People* barely acknowledges the PCB threat in mother's milk while endorsing breast-feeding. Chapters on healthy infants, pregnancy, and infant care make no mention of the trade-offs and risks of breast-feeding.

Only in a chapter on toxic substances is there a brief description of the PCB problem. It concludes... "PCBs have been found in tissues of humans and in the milk of nursing mothers. When fed experimentally to nursing monkeys, PCBs have led to serious injury of offspring." Period. No recommendations. No advisories.

When contacted, a spokesperson in the Surgeon General's Office said that stronger warnings were not appropriate until other agencies in the government take a firm position on the matter.

We suggest, instead, that the risks be aired and serious studies begin immediately.

The Benefits of Breast Milk

During the 1970s, scientists undertook new studies to better understand the attributes of breast milk. They discovered that mother's milk contains:

• a thyroid hormone which babies cannot produce for themselves;

• adequate amounts of iron and zinc, long thought to be deficient in breast milk;

• antibodies and white cells to fend off disease.

For the mother, nursing immediately after birth acts as an anti-hemorrhaging agent while the uterus contracts. Nor is any drug, such as DES (a carcinogen), necessary to relieve the discomfort of breasts temporarily engorged with milk.

Of all the benefits claimed for breast-feeding, the ones least susceptible to measurement are the psychological and emotional. Through skin contact and sensory stimulation, a woman who enjoys nursing may be passing on a real sense of security and attachment to her baby. No one can say for sure how valuable this is for the mother-child relationship.

What Kind of Balance Sheet?

One person who has consistently focused on both the risks and benefits of breast-feeding is Stephanie Harris, a chemist by training, who worked on pesticide issues for Ralph Nader's Health Research Group. Aware of the generally high levels of pesticides in animal products, Stephanie became a vegetarian three years before the birth of her first child. She grows most of her own food organically, filters her water, and takes careful steps to avoid toxic chemicals. Due to a family history of allergies, she chose to nurse both her children, though she knows her milk contains some PCBs.

Stephanie Harris decries the inadequate information available to pregnant mothers about the poisons in their bodies. It was her persistence and concern that helped force the issue before the regulators in 1976. Then, as now, Harris says, there is no inter-agency governmental group responsible for both the monitoring and analysis of the PCB threat in breast milk.

"It is an issue that slips through the cracks," Harris commented sadly.

Meanwhile, parents like Stephanie Harris are faced with difficult choices. She offers women who may become pregnant these precautions for limiting an infant's exposure to PCBs:

• Eliminate from the diet freshwater fish, like salmon and carp, and bottom-feeding estuarine fish, like catfish, flounder, and sole.

• Eat more fresh vegetables, especially organic vegetables, and grain products since more PCBs are ingested through animal and dairy products.

Note: don't expect diet to make a difference overnight. Studies show that PCBs remain in the body for years after the initial exposure. All that you can do is stop increasing the PCBs to your system.

• Lose weight before pregnancy to reduce the fatty tissue that keeps PCBs in the body. Once pregnant, maintain your weight. Weight reduction during the pregnancy only releases more contaminants from the fat cells into the blood stream and eventually through the placenta.

• Avoid breast-feeding if you know you are exposed to unusually high levels of PCBs where you work or live.

• Consider breast-feeding one or two weeks to reap the known benefits of colostrum, the secretion preceding milk, then shift to bottle-feeding.

• Mix breast and bottle-feeding throughout the lactation period depending on the amount of PCBs you suspect yourself of carrying.

(Unfortunately, no inexpensive and reliable tests exist for detecting PCBs in mother's milk.)

Stephanie Harris sums up the dilemma best:

"The choice [a mother makes] is based upon speculation and hypothesis rather than established scientific fact. It is rare that a clear 'right' decision exists... Yet the continuing health and well-being of the infant hangs in the balance."

Suggested Readings

Allen, J.R. et al., **Fd. Cosmet. Toxicol. 15**, 401-410, 1977.

Goldman, A.S. et al., **J. Pediatrics 82**, 1082-1090, 1973.

Harris, S.G. and Highland, J.H., **Birthright Denied: The Risks and Benefits of Breast-feeding.** Environmental Defense Fund, 1977.

Harris, S.G., **Strategies for Public Health,** ed. Lorenz, N.S. and Devra Lee Davis, Van Nostrand Reinhold, N.Y. 1980.

Kodama, H. et al., **Arch. Env. Health 35**, 95-100. 1980.

Rogan, W.J. et al., **The New England Journal of Medicine 302**, 1450-1453. 1980.

CSPI Calls for Action

Breast-feeding infants with milk contaminated by PCBs poses an acute dilemma for new parents and a generalized risk for society-at-large. The Center for Science in the Public Interest urges:

• The National Institutes of Health to hold conferences and workshops aimed at identifying the research need in this little-understood field. We would expect Congress and NIH to follow through, with funds to commission such research.

• A national conference sponsored by a consortium of governmental agencies to educate the public about the long- term implications of PCB contamination.

• The Institute of Medicine of the National Academy of Sciences to undertake a study weighing risks and benefits of both breast-feeding and bottle-feeding.

The Freshness Illusion:

A Few Good Words for Frozen Food and Some Serious New Questions About the Content of Supermarket Produce

BY BRYAN JAY BASHIN

HARROWSMITH, JAN/FEB 1987, P. 41

The availability of fresh produce year-round has to rank as one of the notable achievements of the late-twentieth century. Because of it, our diets have probably changed more in the last 10 years than in the entire preceding 100. In winter, we are no longer obliged to choke down mushy canned beans, to thaw an icy green block that once was peas, are no longer limited to the diet of our forebears: the cabbage, root crops and squashes that endure in storage. Shiny fruits and crisp vegetables entice us as they sit on their shelves in supermarkets across the North. Spurred by the effusions of the popular press and a lean and supple clique of aerobic mannequins in designer leotards, Americans have embraced the notion of freshness in a way that borders on faith.

Misdirected faith is not without its dangers, however. The problem with produce in winter is that, for most North Americans, it simply cannot be fresh. It can be uncooked — sometimes it is even unprocessed — but never is it fresh. Look at the produce section of your local supermarket: what you get is not what you think you see.

This manna that is displayed daily on supermarket shelves has been bred for durability, picked green and waxed for show. It may be old enough to grow whiskers and have put on more miles than an Airstream trailer in summertime. It can be expensive, and by the time it arrives at the supermarket, its nutritional qualities are in decline. And for those who reflect upon their actions, the implications of buying fresh can be troublesome indeed, as, increasingly, our winter produce come from foreign lands, where farm workers labor under conditions so exploitive and inhumane that you probably do not want to know about them. But nonetheless, those fruits and vegetables sitting on ice are always bought with the belief that they are somehow better for us than processed food and that they enhance our "lifestyle."

The Boom

Grocers have never seen anything like it. On average, an American eats 26 pounds more fresh fruit and vegetables per year than he or she did 10 years ago. Today, the typical produce section stocks more than 250 items, five times the number of 10 years ago, and space devoted to "fresh" has more than doubled.

Other statistics detail the scope of the boom in fresh food. Since 1976, broccoli consumption is up 169 percent, grapes 117 percent, cauliflower 100 percent, spinach 80 percent, nectarines 71 percent, strawberries 41 percent, cucumbers 38 percent, green peppers 24 percent.

The average American now eats 143.2 pounds annually of 11 fresh vegetables tracked by the United States Department of Agriculture. This feast includes 52.1 pounds of potatoes, 25.5 pounds of lettuce, 19.4 pounds of onions, 15.8 pounds of tomatoes, 7.7 pounds of sweet corn, 7.6 of carrots, 7.4 of celery, 2.9 of broccoli, 2.2 of cauliflower, 2.1 of honeydew melon and .5 of asparagus. (The USDA does not keep statistics for cabbage, green beans and peas.)

The consumption of fresh fruit is also up, to 91.1 pounds, according to the USDA. The average diet includes 25.6 pounds of bananas, 18.1 of apples, 12.4 of oranges, 6.9 of grapes, 5.7 of grapefruits, 4 of peaches, 3.1 of strawberries, 2.8 of pears, 2.4 of lemons and 1.5 of pineapples.

By contrast, we eat only 60 pounds of frozen vegetables – more than two thirds of which are potatoes – and between 1967 and 1983, canned fruit consumption dropped 40 percent, and frozen fruit sales dropped 20 percent. In some areas, the economic impact has been enormous: more than 15 major canneries have closed in California in the last decade, eliminating more than 17,000 jobs.

Annually, sales of fresh produce have reached $35.9 billion, convincing economic testimony to its perceived value.

Just How Good is Fresh?

Consumers often pay premium prices for out-of-season fresh food, and even where price differences between fresh and frozen are slight, the price per pound for fresh food is deceptive, since the consumer pays for that which is later discarded during preparation.

Indeed, many nutritionists say that frozen vegetables are a better buy than the allegedly fresh foods available at the supermarket. In its just released *Status Summary on the Effects of Food Processing on Nutritive Values,* the 34,000-member Institute of Food Technologists looked at fresh versus frozen foods and concluded: "Foods which are processed with good manufacturing practices from high-quality, freshly harvested 'garden-fresh' commodities will frequently have a higher nutrient content than freshly harvested 'market-fresh' commodities which have been improperly handled during transportation and/or stored for a few days or more."

The Institute of Food Technologists also reported that natural variations in vitamins and minerals in raw food are far greater than the variations between raw and processed foods that are caused by the processing. The Institute reports that raw foods can gain or lose nutrition based on "genetic differences, climatic or soil conditions, maturity at harvest and handling conditions following harvest." Samples of fresh tomatoes and carrots tested by the Institute ranged from half to twice the expected levels of vitamins C and A.

A look at some other vegetable facts might further compel food shoppers to rethink their habits.

• By the time asparagus travels the 2,700 miles from California to New York, even by refrigerated truck, it has lost two-thirds of its vitamin C. (Assessing a vegetable's remaining level of vitamin C is often a good indicator of its overall quality.)
• Refrigerated "fresh" broccoli will have lost 19 percent of its vitamin C in 24 hours, 34 percent in two days.
• If fresh, frozen and canned peas are cooked the same way, they will have almost the same amount of vitamin C.
• Although baby vegetables like carrots and potatoes are in vogue – and pricey – they are less nutritious than their mature counterparts.
• Peeling the skin of a cucumber eliminates almost all of the vegetable's vitamin A.
• Unless you intend to eat delicate foods like tomatoes, berries and leafy vegetables within two days of buying them, you would do better from a nutritional standpoint to buy frozen. And you would be likely to save money.

According to USDA nutritionist Ruth Matthews, "A busy housewife who only goes to the store once a week, if she has a good freezer, she might be better off buying frozen."

Adel Kader, a food scientist at the University of California at Davis, agrees: "If you compare frozen beans with fresh shipped to the Northeast, it is likely that the frozen food would actually have higher nutritional value."

David Reid, another food scientist at UC-Davis, recalls the time he and one of his graduate students decided to do some basic chemical testing. They went to the local supermarket and bought Brussels sprouts and cabbage. Both are considered good sources of vitamin C: straight from the garden, one crop of Brussels sprouts should provide slightly more vitamin C than a cup of fresh orange juice, cabbage about one-third that amount.

"We brought the vegetables to the lab and started to run tests, but we couldn't find any measurable amounts of vitamin C," Reid says. "Buying that supposedly fresh stuff out of season, it had probably been stored for months, and it showed."

Long-distance shipping is one of the main reasons why "fresh" fruits and vegetables usually are not. A United States Department of Defense report estimated that the average distance the nation's food travels before being eaten is 1,300 miles.

That journey subjects produce to various traumas of the road, which speed its nutritional devaluation. "Folic acid [one of the B vitamins], ascorbic acid [vitamin C] and other vitamins are easily lost if the product is bruised or wilted or mishandled," says USDA nutritionist Matthews, who has studied nutrient loss in fresh vegetables. Matthews also points out that by cutting melons, squashes and other vegetables in half, supermarkets may render them nutritionally inferior to their frozen or canned counterparts. "Once the plant material is cut," Matthews explains, "then the enzymes begin to cause oxidation and a rather rapid loss of vitamin B, folic acid, vitamin C and some vitamin A." Nonetheless, pre-cut fresh produce carrying major food companies' brand names may be the next addition to the produce shelves. Kraft, Pillsbury and Dole are all hoping to launch such marketing balloons on the hot air that now surrounds fresh produce.

Besides nutrition, flavor is the other argument for buying fresh produce. As many people know by now, foods for the produce section are picked green. The produce has to be harvested when it is still firm enough for transportation and young enough to keep its natural enzymes from causing spoilage before it arrives at the market. Before being shipped, some fruits and vegetables are gassed with ethylene, which produces ripe-looking color and texture. Although the fruit or vegetable may appear ripe, it will never develop the natural sugars that distinguish vine-ripened produce.

"Really fresh food has sugars and acids and volatiles," notes UC-Davis' Kader. Its aroma and flavor are distinctive. But as Kader says, "Anyone with a garden knows the difference between tomatoes from the backyard and tomatoes from the market."

The Big Chill

Even if the most nutritious alternative to fresh produce is frozen, frozen food can deteriorate after processing if allowed to warm, say, on a loading dock or in the back of a station wagon. With frozen vegetables, "warm" is a relative term. Spinach, even held at zero degrees, loses 80 percent of its vitamin C in a year; at the same temperature, lima beans lose half their vitamin C, 45 percent of their riboflavin and 26 percent of their niacin.

Food technologists have developed an easy rule of thumb to compute these losses: for every five degrees above zero at which frozen produce is stored, its nutritive life is cut in half. At 15 degrees F, spinach would lose 80 percent of its vitamin C in six weeks.

Finally, some food scientists would say that the emphasis on fresh produce misses the essential nutritional point. What the American diet lacks is not usually vitamins, but minerals and fiber. Despite all the hoopla about fresh foods, there is no difference between fresh and frozen on that score.

The best fruits and vegetables are grown locally, but the trend toward fresh produce has, paradoxically, hurt local growers. "The distribution system discourages local growers, because a lot of big purchasers won't buy from them," says Ellen Pahl of Ag-Market Search, a nonprofit project designed to encourage farmers to grow produce for local markets. "The purchasers depend on year-round suppliers who depend on year-round contracts. But we think that if we can get local contracts going, purchasers can buy locally and maintain their other sources when the local produce isn't in season."

Most areas have become dependent on growers outside their immediate regions. Philadelphia, which used to be surrounded by rich farmland that once supplied most of the city's food, today has 70 percent of its food trucked in from out of state. Iowa, in the heart of farm country, imports 91 percent of its fruits and vegetables.

Home Ec 101
A short course in healthy cooking

Given the nutritional uncertainties of fresh produce, what is the food shopper to do? The first thing is to be a careful cook, because the greatest losses in the nutritional content of foods happen not in distribution but during cooking in the home. According to the new Status Summary on the Effects of Food Processing on Nutritive Values released by the Institute of Food Technologists: "As a result of the large vitamin losses that occur during cooking in the home, the actual vitamin content of table-ready foods is frequently about the same, regardless of the type of processing, or lack of processing, the food has undergone."

So here are a few suggestions for how consumers can reduce vitamin and mineral losses during cooking, along with a few shopping tips.

• Once foods are cut, oxidation and deterioration of vitamin C is rapid, so wait until the last minute to cut into vegetables and fruits. Acids, like lemon juice, can be applied to the cut surfaces of any vegetables to preserve their vitamin C. When making coleslaw, for example, the quick addition of an acid dressing will stop vitamin C losses, or tomatoes can be added to cut zucchini to help preserve the squash's vitamins.

• Use absolutely as little water as possible when boiling vegetables. Vitamin C dissolves in water and nutrients are leached out, particularly when vegetables are cut up. Boiling carrots, for example, will leach out a third of their potassium; steaming them will preserve all of their minerals.

• Bring cooking water to a boil, then throw in the vegetables. Allowing water to heat up slowly with the vegetables in it leaches nutrients and destroys vitamins.

• Vitamin A dissolves in fat, so if you cook vegetables in butter or oil, vitamin A will accumulate in the cooking fat. Reuse that rather than new butter or margarine.

• Cook in covered pans to shorten cooking time and decrease the amount of water needed.

• Leafy vegetables, which have a large surface area, lose the most nutrition, so the less cooking the better.

• You will lose fewer nutrients if you cook vegetables whole and cut them up after they are cooked.

• When storing food, for maximum vitamin retention, protect them from heat, air and light.

• Keep your home freezer below zero degrees F; 10 degrees below zero is preferable. For every five degrees colder you keep your freezer, the nutritional life of your vegetables will be doubled.

• When buying frozen vegetables, make sure that they rattle in the box or bag. Vegetables that have thawed and refrozen will stick together in a clump.

• David Reid, a professor of cryoscience (freezing technology) advises: "Go into a supermarket and squeeze the ice cream. If it is soft, go to another supermarket to buy your vegetables." Reid, who lives in the heart of California's productive Central Valley, says of his family's shopping habits: "We eat whatever is the least expensive. Once the vegetables come down the distribution chain, there's not much nutritional difference between fresh and frozen."

If consumers were to differentiate between local produce and that imported from outside the region, substantial dollar and energy savings would result. A four-day journey from California to the East Coast adds between 6 and 15 cents per pound to the price of produce. It cost $2,000 more per truckload to ship broccoli east from California than to truck it locally, and annually, residents of New York City pay $6 million in trucking costs for their broccoli. Generally speaking, for every $2 spent on growing food, another $1 is spent on moving it.

A return to local crops would also counter breeding of varieties especially for interstate commerce. For example, today only two kinds of peas account for 96 percent of the United States' commercial production.

The development of "shippable" fruits noted for their "structural integrity" may be approaching its zenith. Take the peach: for 5,000 years it took on that special blush only when it grew ripe. Appearing today on supermarket shelves are several new varieties bred to ship well. They develop that seductive peach-colored hue while still immature and crunchy inside. If Madison Avenue had developed this fruit, we would call it deceptive advertising. In the produce section, the word is *caveat emptor:* let the buyer beware.

That Car Wax Shine

One unwanted bonus consumers get when they buy fresh foods is the waxy coatings used on them–and not just the obvious waxing of rutabagas and cucumbers. Fresh produce may be coated with a film that food industry groups say is needed because it cuts moisture loss by 30 to 40 percent, preventing the shriveling of produce and the onset of decay, and thereby extending shelf life.

In this country, many of the major fruits and vegetables are coated, including apples, avocados, bell peppers, cantaloupes, cucumbers, eggplants, grapefruits, lemons, limes, melons, oranges, parsnips, passion fruits, peaches, pineapples, pumpkins, rutabagas, squashes, sweet potatoes and in some growing areas, tomatoes and turnips.

"Many consumers are unaware of the number of fruits and vegetables that are waxed," says Joe M. Regenstein, associate professor of food science at Cornell University.

"Waxing is quite common with cucumbers, peppers and certainly tomatoes, to name a few," adds Robert Kasmire, a post-harvest specialist at UC-Davis. "People buy with their eyes," he says by way of explanation.

The FDA has approved a half-dozen waxes to be applied to some 18 varieties of fresh fruits and vegetables. Some floor and car waxes are largely based on the same ingredients. The approved waxes include: petroleum-derived paraffin, palm-derived carnauba wax (also found in both floor and car waxes) and candelilla (used in furniture waxes), which is produced from a reedlike plant. Shellac may also be used, and the FDA has approved a synthetic petroleum-derived waxy coating made of polyethylene. Also widely used is oleic acid, a waxy substance that can be made from either animal tallow or vegetable oils or synthesized from petroleum. Finally, the FDA has approved a new class of waxes that can be manufactured from animal tallow and that are used in some floor waxes.

"They don't need to use the waxes," contends Adel Kader at UC-Davis. "Most of the need is cosmetic. I suspect if consumers would object strongly enough, the shippers would stop using waxes immediately."

"Waxing is just an embalming technique," says Steve Pavich, a large California organic grower who does not wax. "It allows the mediocre grower to do a mediocre job growing it and then to pretty it up – and the consumer doesn't know the difference."

To remove the waxes, you must wash the produce in detergent – rinsing with plain water does no good.

Much of the recent concern about waxes centers on those that can be made from animal tallow. Such waxes are unacceptable to a variety of religious groups and individuals: Orthodox Jews, Moslems, Hindus, Seventh-Day Adventists and vegetarians. To inform prospective buyers, Cornell's Regenstein has proposed a simple system of three classifications: "unwaxed," "waxed without animal derivatives," and simply "waxed."

But Sharon Bomer, director of government relations for the United Fresh Fruit and Vegetable Association, thinks any labeling system would be unwieldy, both for producers and packers. She says that it would also be burdensome on retailers to have to affix a variety of tags on, say, Red Delicious apples, which they now combine in a single bin. "Under the proposal, a retailer might have to maintain a half-dozen separate bins of waxed Red Delicious apples."

Regenstein says that although major manufacturers of coatings have informally agreed not to use animal-based waxes, he remains concerned about the waxing of foreign produce, which makes up a growing percentage of fruits and vegetables sold in the U.S. He thinks consumers should be able to know when they are eating waxed food and what the wax contains.

Supposedly, federal law is on his side. According to FDA regulations, waxed produce must be labeled as such. With bulk items, the retailer is required to affix a card next to the price of the produce, indicating the type and composition of the wax used, or the seller may display the original shipping carton alongside the produce. The law also requires labeling of any fungicides or fumigants used on the produce.

Although the labeling requirement is explicit,

Regenstein says, "Nobody I know has ever seen an ingredient statement on fruits or vegetables." In other words, supermarkets routinely scoff at the law when they display produce.

In a June 1986 letter to Mark Brian, a California consumer activist, FDA consumer affairs officer Gertrude Gabuten explained the FDA's problem: "Although the federal law requires labeling of fresh fruits and vegetables at the retail level, we have never had the manpower or resources to inspect the thousands of food and other retail stores for possible labeling violations. Therefore, we have traditionally relied on state and local agencies to enforce labeling requirements for retail displays of bulk produce."

Citrus and Sheep Wormer

It is rare that consumers even notice that their produce is waxed, but if they were to read the packing labels on produce cartons, they would also learn about the pesticides applied after the harvest. Shippers will often mix the wax coating with sprouting inhibitors and fungicides to slow decay. Orthophenylphenol may be mixed with the wax on tomatoes, and thiabendazole and diphenyl are commonly used with fruits. To anyone who has raised livestock, thiabendazole will sound familiar – it is commonly used as a medicine for worming sheep.

"There isn't a banana or pineapple in any fruit bowl in this country that doesn't have thiabendazole in it," Mark Brian insists. "Nobody knows it exists, though, because the supermarkets destroy every trace of packaging that has information about it."

Speaking for the fruit and vegetable industry, Bomer downplays the significance of the unlabeled pesticides on fresh produce. "When you think of chemicals, you think of them as unsafe," she says. "It is a perception. In reality, they have been tested." Tests show that as a group, fungicides are worrisome, many of them being oncogenic – likely to produce tumors and possibly cancers. Although they are widely approved for use, a major report on the public health consequences of current pesticide regulations is due out within a couple of months. In its draft form, that study, being done under the auspices of the National Academy of Sciences, is said to give considerable attention to fungicides.

Yet Bomer's organization still resists labeling. "If it were a safety question, it would be another matter," she says, "but since all waxes are edible, we find this an unnecessary concern. We're opposed to any labeling of fresh fruits and vegetables because it is impossible to do on the retail level."

Who is Minding the Store?

The increased consumption of fresh produce brings with it a greater possible risk of exposure not only to post-harvest treatments, but also to farm chemicals, as vegetables and fruits come to the table with less processing, boiling and blanching.

Charles M. Benbrook, executive director of the board on agriculture of the National Academy of Sci-

---

The NutriClean Seal of Approval
For fresh produce you can trust

Finally, someone is making it possible for the consumer to know precisely what nutrients and pesticides are in a particular fruit or vegetable. Ohlone, Inc., of Oakland, California, is working with growers to have the company test their produce, and the results, which will be printed on cards that accompany the produce on supermarket shelves, will be available for all food shoppers to see. Linda Brown, spokesperson for Ohlone, says, "We're trying to establish that the safety of produce can be a given. There is plenty of good food being grown, and not only is it safe, it is nutritious."

Ohlone, which is subcontracting the actual testing, has developed Silver and Gold Seals to distinguish safe and nutritious produce. To receive a Silver Seal, produce must not exceed one-tenth of the Environmental Protection Agency's allowable pesticide residue levels and must have at least 75 percent of the levels of key nutrients that the USDA says would be available in that specific fruit or vegetable.

The Gold Seal is based on more exacting standards. It requires no detectable pesticide residues either in the food or in the soil and water used to grow the food, and the rating of key nutrients must be 100 percent of the expected level. In addition, no category one or category two chemicals – the most toxic, like paraquat, Roundup and parathion – may be applied during the growing and harvesting of the produce. To qualify for the program, growers must fully disclose all chemicals that are applied on their fields.

The Ohlone program began last spring, under the NutriClean label, in a select few supermarkets in Los Angeles and San Francisco. So far, NutriClean seals have appeared on apples, grapes, potatoes, lettuces, peaches, nectarines and peas, and the company is working with growers to expand the number of participants. The company also plans to establish a network of authorized distributors, who will be required to maintain Ohlone's standards of refrigeration, shipping and storage.

So, soon, it will be possible for you to know what you are getting when you buy produce at the supermarket.

ences' National Research Council, says there are about 150 chemicals that regularly appear as residue in food, and something like two dozen appear at levels of "toxicological significance." Benbrook points out that the problem of dealing with chemicals in fresh food occurs not because of the sheer variety of substances, but because of the slow pace of U.S. regulators. Though there are hundreds of dangerous pesticides to monitor, on the bulk of food products, relatively few are cause for concern, according to Benbrook. For example, just six fungicides account for 90 percent of all fungicides used on tomatoes. "A lot of people point to the figure of 45,000 chemicals used in agriculture," he said. "That makes the problem of pesticide residue in food seem more daunting than it actually is. If one adopts a more sophisticated view, it is possible in the next several years that we can move to regulate those pesticides that are most significant."

Benbrook says that his major concern is that under the current administration, the pace of regulation is so slow that judgments are made on only two or three chemicals a year. "The big problem in this country and in Canada is that there simply isn't a strong constituency pressuring for food free of residues of widely used pesticides. Perhaps the day will come when people will raise such pressures, but it has not happened yet – even in the face of clear information that there are residues in the food supply."

Northern consumers hoping to eat their way into health by buying out-of-state fresh produce from places like Florida and Mexico may want to think again. Many of the humid southern regions that grow these crops have a distinct problem with mold and fungus and consequently need to use more fungicides on them. "The more humid an area," Benbrook says, "the more reliant on fungicides, which tend to be the most problematic class of pesticides. Florida peppers and melons, for example, are reliant on substantially more pesticide treatments; in contrast, Texas and California produce uses less because of the low humidity and climate. There are many fresh fruit and vegetable crops in Florida that are sprayed 20 or more times a year with five or more active ingredients."

One recent study found traces of 19 different chemicals in roughly half of the produce sampled, including up to four types of pesticides on a single strawberry. Eight of the nineteen pesticides found in the vegetables are known or suspected carcinogens.

Even Adel Kader, who says he has no concern about eating fruit coated with a fungicidally treated wax, remarks: "History has taught us to be cautious."

The nation's principal guardian against the appearance of pesticides in produce is the FDA. Through its system of 19 regional laboratories, the FDA tests thousands of samples of foods to be sure they are safe. But last September, Congress' investigative arm, the

General Accounting Office (GAO), released a report sharply critical of the FDA's speed in analyzing food samples. The GAO found that the regional labs were so slow at testing products that adulterated and misbranded food was routinely reaching the marketplace.

After looking at 82,000 samples tested by the FDA between 1983 and 1985, GAO's analysis showed that the 19 field laboratories took an average of 28 calendar days to complete "product processing." The labs' efficiencies varied widely. One laboratory completed its tests in an average of as little as 8.5 days while another averaged a glacial 87.6 days.

That was not ordinary testing. That was urgent testing on product samples for which the FDA had some reason to suspect that the food had been tainted – such as consumer complaints or a record of poor compliance. But what good is a report on a shipment of produce one month after it has arrived at the supermarket?

The GAO report was unqualified in its criticism of the FDA's food-testing procedures: "*In a current review of pesticide residues in foods, we found that the FDA was not able to prevent any violative products from reaching the consumer because of untimely laboratory processing* [author's emphasis]. By the time the FDA identified the products as violative, they were often no longer available because most agricultural products are perishable and therefore move rapidly from farms to consumers."

On routine surveillance, which turns up a quarter of all violations, lab results were even slower to appear, 38.3 days on average in the first half of 1985. While San Francisco needed an average of only 6.1 days to process items, New York's lab on average took an appalling 164.9 days to process its samples. On imported items, the GAO said the FDA took an average of 8.1 workdays to process samples, though some labs like Chicago's took nearly 20.

These long delays are not a result of complex testing procedures. Of the 28 days it took the FDA on average to test a product, nearly two weeks were wasted as the sample simply sat in inventory waiting to be analyzed. At one lab, samples sat an average of seven weeks. At that rate, FDA scientists were not protecting consumers – they were doing archaeology.

The FDA's inadequacy as a guardian of safe and healthy food becomes a more significant issue now that, increasingly, our food comes from foreign countries with safety standards about which the kindest characterization one can offer is that they are lax. Safe importing of produce is not a mere hypothetical concern, as the case of the Chilean strawberries makes clear. Several years ago, pending testing, 4,465 pints of Chilean strawberries were allowed to enter the country on November 30. The laboratory received samples for testing on December 5, but did not complete the work until the following February. The tests revealed excessive amounts of chlorothalonil, an unsafe pesticide, but the FDA took no action, assuming that after two months, there would not be any strawberries left to act on.

Mexico's Sorry Tomatoes

Boosters of fresh produce have treated their subject with a light, bubbly prose akin to the sparkling water they favor, ignoring incidents like the Chilean strawberry caper. The fundamental shift in the source of our winter produce cannot remain overlooked, however, if consumers are genuine about their concern for their own physical welfare and if they have the simple decency not to ask the people who grow their food to jeopardize their lives.

Mexico alone grows 70 percent of the vegetables consumed in the United States during January and February; 50 percent of the produce sold in supermarkets between December and April is imported. All 11 billion bananas we eat annually in the continental United States are imported; one-third of the grapes we eat come from Chile. The steady rise in food importation has reached the point where, in 1986, for the first time in 27 years, the nation brought in more food than it sent abroad. Today's imported food bill, at $24 billion, is twice what is was as recently as 1980.

It is true that imported produce is cheaper than U.S.-grown. Mexico produces tomatoes for half of what it costs California. Chilean agribusinesses pay their field hands $1 an hour or less; Californian farm workers earn up to $10 an hour. *New York Times* agriculture reporter Keith Schneider writes: "It is possible to fly a load of apples all the way from Chile to Boston and sell them for less than apples grown in New York State."

We are, in effect, willing to pay transportation companies – but not our farmers – for our food.

In San Diego County, California, 5,300 acres of tomatoes were planted as recently as 1980; last fall, only 2,000 acres were planted. On the Baja peninsula of Mexico, 15,000 acres have recently been planted in tomatoes, and American dollars are heading south, chasing cheap produce. Twenty years ago, Mexican food imports cost us $29 million; today, they exceed half a billion dollars.

(It is worth noting that the importation of produce includes frozen food as well as fresh. During the last three years, for example, Pillsbury's Green Giant started freezing cauliflower and broccoli in Mexico; today, their factories freeze a third of all the cauliflower and broccoli Green Giant packs. Pillsbury can freeze and ship Mexican broccoli to the U.S. for 35 cents a pound versus 45 cents a pound for California broccoli.)

Now, though, critics of the binge in imported produce are raising questions about how safe that produce is to eat and how ethical it is for Americans to buy a product that is grown by an inhumane and exploitive farm-labor system. What are the costs of eating foreign produce?

Exhale When They Spray

In the Mexican state of Sinaloa lies the Culiacan valley, a region that began a huge boom in fruits and vegetables in the 1970s. Today, the valley's half-million acres of irrigated land are so productive that the region grows between one-third and one-half of all fresh tomatoes sold in the U.S. from December to May. Two-thirds of all Mexico's U.S. exports of tomato products come from here, as do vast quantities of bell peppers, cucumbers, eggplants, zucchini, summer squashes, onions, garlic and chilies.

But there is trouble in this valley. Increasingly, reports are reaching the U.S. of pesticide poisonings and squalid living conditions. "A living hell," said one researcher.

Angus Wright, professor of environmental studies at California State University in Sacramento, has repeatedly journeyed to Culiacan in an effort to document conditions there. He says that the predominant chemicals used to spray fields in Culiacan today are organophosphates. Unlike older pesticides like DDT, the organophosphates are less likely to persist in the environment and accumulate in the food chain. The trade-off is that they are ever so much more acutely toxic and are readily absorbed through human skin. Wright has observed more than 50 pesticide applications by airplanes, backpack sprayers and tractor-drawn equipment in the Culiacan fields. He writes in his report *Rethinking the Circle of Poison*: "I noted a virtually complete absence of protective gear. There were no rubber boots, no rubber aprons or coveralls, and no respirators or face masks...often applicators wore short-sleeved shirts and they often wore short pants. Cloth running shoes, open sandals or rubber thongs were the most common footwear. Some applicators and mixers were barefooted."

Wright goes on to describe living conditions: families use discarded pesticide containers for household purposes, and fieldworkers live in open sheds next to fields that are regularly sprayed by planes.

He writes: "The fumigation aircraft and backpack spray crews must spray to the edge of the fields, meaning that they must spray to the edge of campamentos, where children play, people cook and eat and where people lie down to sleep at the end of the day... most still draw their water from nearby irrigation ditches."

He also reports: "I observed fumigation airplanes spraying directly over crews of workers, composed of 20 to 30 men, women and children." Wright recalls a conversation with an employee of the Culiacan Growers' Association who told him: "We try to tell them [workers] how to protect themselves, to wear the right clothes, to wash the tomatoes before they eat them, and to exhale when the airplane passes over them while spraying."

The spraying rates are among the highest in the world. From January to May, fields are sprayed every three or four days, and crops may be sprayed between 25 and 50 times during the growing season.

After searching through government records, Wright reported: "In four years, health authorities in the municipio of Apatzingan, Michoacan, reported more than 1000 cases of acute pesticide poisoning...We estimate that the number of pesticide poisoning cases diagnosed in clinics in the Culiacan valley during the months of heavy spraying from January through May is within the range of 40 to 60 cases per month."

Wright comments, "To give you an idea how widespread the cynicism is, in Baja, Mexico, the foremen are being given boxes of atropine injections to shoot up workers in the field at their discretion." Atropine, a belladonna derivative, is an antidote to the symptoms of pesticide poisoning.

To keep the number of reported poisonings down, Wright says, growers send workers to private doctors where statistics are impossible to obtain. The workers live in fear of reprisals if they complain or even if they seek treatment at government clinics. "The fact that many foremen travel about the fields in the company of a pistolero, a gun thug, does nothing to allay such fears," Wright remarks.

What is happening in Mexico today is different from the situation of five years ago, when illegal and dangerous pesticides were shipped from the U.S. to Third World countries, which then used them on produce later sold to the U.S. Observers agree that today nearly all of the 14,713 tons of pesticides used in Mexico are manufactured there and registered for use by the EPA. At the same time, to meet U.S. cosmetic standards, the Mexican farmers are applying pesticides at high rates.

Two government agencies oversee the quality of imported food. The USDA inspects every shipment of Mexican produce coming into the country to make sure it meets grading standards and to check for insects and disease. The FDA, which is responsible for the monitoring of toxic residues, checks fewer than one in ten shipments of imported food for pesticide residues, despite the heavy rates at which pesticides are applied, and the number of FDA inspectors nationwide has stayed constant during the past five years.

Diane Baxter, staff scientist at the National Coalition Against the Misuse of Pesticides, notes, in addition: "When the FDA inspects imported foods, they test for only about one-third (148 of 312) of known pesticides registered for use on food crops in the U.S. And there are unregistered chemicals, which can go undetected."

But some within the food industry do not blame the government regulators. They say the fault lies with consumers. As Howard Weatherspoon of the Frozen Vegetable Council says: "If they eat seven-day-old pesticide-laced Mexican green beans, they deserve what they get."

The last two times Angus Wright visited Mexico, he became physically sick from pesticides while witnessing applications of parathion. He has learned firsthand the living and working conditions that currently provide us with the wintertime luxury of fresh food, and he refuses to put money into the pockets of Mexican growers. "When I walk into a supermarket and see the tomatoes that come from Culiacan, I get nauseous and angry, and I can't even look at them. It is so violent and brutal there, the most incredible kind of exploitation.

"I try to buy in season and as close to home as possible," Wright adds. For those who do not live in a warmer climate, he advises: "I would buy the crops that can be grown late in the season or preserved and kept, such as lots of hardy greens and cole crops – the kind of thing that Europeans ate for centuries. Root crops and winter squash are good too." In effect, Wright suggests that we return to the diet that preceded chemical agriculture, the diet that sustained many of our ancestors for generations. Or as Wright calls it "Our soul food."

How Bugs and Pests Rout the Chemists

THE ECONOMIST
MARCH 21, 1987, P. 97-98.

The ingenuity of chemists is no match for the versatility of evolution. In the battle against plagues and pests, the enemy has for some time deployed a strategic defense initiative resistance. Bacteria are evolving resistance to antibiotics and insects resistance to pesticides faster than new chemicals are being invented. Before long, at this rate, medicine will revert to the days before penicillin, and agriculture to the days before DDT. Little prospect of relief is in sight. Ways exist to avoid resistance but chemical companies, doctors and farmers have no incentive to adopt them.

Resistance is one of the clearest instances of Darwinian natural selection. As soon as penicillin is injected into your body, an advantage accrues to any rare bacterium that happens to possess a way of disarming the chemical. It survives while the others perish. Resistance then spreads from bug to bug because most of the resistance genes lodge themselves in plasmids: small loops of DNA that bacteria swap with each other. Resistance is usually irreversible. It can also move between species of bacteria: a harmless soil organism made resistant by antibiotics on a farm can pass the resistant plasmid to a strain of Salmonella. In the United States, half of all antibiotics are used to make farm animals grow faster.

Again for Darwinian reasons, the enemy gets better at developing resistance: selection encourages the quick evolvers. Five successive generations of pesticides have passed since DDT was first used in the 1940s. The time it takes for the number of species of insects resistant to each new generation to double has fallen from 6.3 years for DDT to one year for the newest generation of pesticides, pyrethroids. Colorado potato beetles have beaten almost everything thrown at them. In the beginning, it took them many years; today, they can beat a new insecticide in a season. Meanwhile, many of the insects that eat Colorado beetles are still susceptible. Spraying simply encourages the beetles: they are, says Dr. Robert Metcalf of the University of Illinois at Champaign-Urbana, out of control.

More than 450 species of insect are now resistant to DDT. The most famous is the mosquito, which turned the World Health Organization's 21-year, $2 billion campaign to stamp out malaria into a rout. For example, in 1963, the program had worked so well that there were only 17 cases of malaria in Sri Lanka. But that year resistance appeared. By 1968, the country had an epidemic of more than 1 million cases.

Antibiotic resistance has yet to produce such a medical disaster but it will do so before long. Dr. Stuart Levy of Tufts University has found that more than half the benign bacteria in human guts in Massachusetts are resistant to at least one antibiotic. Benign bacteria can pass on resistance to harmful ones. In the United States, the first of penicillin-resistant gonorrhea appeared in 1976: by 1985, 8,500 cases were reported.

Multiple resistance is a new worry. Dr. Levy has done experiments raising chickens on tetracycline. Within 12 weeks, their gut bacteria were resistant not only to tetracyline, but to several other antibiotics as well. In Tanzania, cholera has evolved resistance to several drugs, though only one was used. The same is happening among insects. In 1976, only seven species were resistant to all five classes of insecticide: now, 17 are, including house-flies, cockroaches and mosquitoes.

Nor is genetic engineering yet much of a help. The first genetically-engineered crops will have genes conferring resistance to herbicides (so that farmers can spray more herbicide on nearby weeds), and resistance to insects conferred by a gene taken from bacteria. Both will simply encourage resistant weeds and bugs to appear quicker.

Cost-ineffective

At the same time the chemists are stumbling. The number of new kinds of pesticides and antibiotics invented each year has fallen fast over the past quarter-century. Between 1980 and 1985, no new pesticides were introduced at all. Meanwhile, costs have climbed relentlessly. Synthetic pyrethroids cost 100 times as much in real terms to develop as did DDT (and the best of them, deltamethrin, 1,300 times as much). The ever more rapid appearance of resistance has put chemical companies in a squeeze. They have less time in which to recoup greater costs, so they encourage wider use, which accelerates the appearance of resistance.

What can be done? The chemical industry may find a way out of its dilemma by developing chemicals that fight resistance. For example, Dr. Samuel Martin and his colleagues at the Walter Reed Army Medical

Center in Washington, D.C. have recently shown that they can reverse resistance in malarial parasites to chloroquine, an antimalarial pill. They did so by learning from cancer cells, which can also develop resistance to drugs. Cancer cells achieve resistance by becoming more efficient at getting rid of the drug and so preventing it accumulating: hence, the cell can resist various kinds of drug after being exposed to only one.

A chemical called verapamil fights this excretion and so reverses resistance. In the laboratory, it also works for chloroquine-resistant malaria. Perhaps, one day, it will be possible to give people penicillin together with a chemical that reverses resistance to penicillin.

Such ideas are futuristic. At the moment, the only answer for both pesticides and antibiotics is to use them sparingly. This not only postpones the spread of resistance, but in some cases can head it off altogether. Antibiotics, Dr. Levy says, should not be used in farms, not be used indiscriminately for weeks before surgery, and should not be available over the counter for all sorts of unsuitable ailments, as they are in many poor countries. In Mexico, scientists have found that the amount of resistance correlates neatly with the availability of antibiotics.

Similarly, insecticides, used carefully, need not produce resistance. In Australia, it is illegal to use synthetic pyrethroids on *Heliothis*, a cotton pest, except during a 40-day period each year when the bug is most vulnerable. After five years of prohibition, no evidence of resistance to synthetic pyrethroids has appeared in Australia. In the rest of the world, cotton growers profligate in their use of the stuff have had no trouble producing resistance.

In another case, Dr. Brian Croft of Oregon State University has persuaded the chemical industry to leave off its label any suggestion that pyrethroids be used in summer against a pest called Psylla. Again, resistance has been avoided. In Italy and East Europe, which have used the pesticide in summer, resistance is rife.

Such practices are hard to introduce. No farmer likes to watch a pest developing when he knows he can kill it, and no salesman wants to tell him to use the stuff sparingly. Dr. Croft admits that some of the best successes owe more to serendipity than to good management. The codling moth, for instance, is not (so far) resistant, while the insects that eat it are. That is pure luck.

Farming will survive – it can breed plants in the right rather than the wrong direction (see box) – but human health will suffer. The only hope is that genetic engineering can deliver weapons that evolve as fast as the enemy. That is not as far-fetched as it sounds: several insects are already fought by spraying them with viruses that give them diseases, and there is talk of doing the same for bacteria. Viruses can out-evolve bacteria, chemists cannot.

Why Fungi Win

Farmers waste time and money spraying crops that hardly need it. In doing so they squander natural disease-resistance in plants and encourage spray-resistance in diseases. This forthright conclusion comes from two plant scientists, Dr. John Barrett, a geneticist at Cambridge University, and Dr. Martin Wolf of the Plant Breeding Institute near Cambridge. They put much of the blame on the practice of a company selling seed for crops with one hand, and chemicals to protect those crops with the other.

Most new varieties of cereal go through a boom-and-bust cycle. At first the new strain is resistant to most of the diseases around. Although some fungus infection does show up, it is not enough seriously to reduce yields. Farmers plant more, and the variety booms. The fungus diseases respond to the challenge and soon new races of the diseases appear that can attack the new strain, which rapidly goes bust.

The cycle can be as short as three to five years. Fungicides could be used to extend the working life of a variety by applying them only towards the end of the cycle. Instead the farmer sees the trivial infection in the first year and, nudged by salesmen, reaches for a fungicide. That gives the fungus a chance to evolve a way to disarm the fungicide. The upshot is that the fungicide becomes useless at the same time as the natural disease-resistance breaks down.

Plant breeders have made matters worse as they move towards hybrids to retain commercial control of good varieties. The breeder grows the patented male and female parents, selling the hybrid seeds. But hybrid crops can be produced only if the female parent has open flowers, to admit pollen. Open flowers also let fungal spores in.

Hundreds of hectares planted with the same variety and treated with the same chemicals pose too easy a problem for pathogens. Dr. Barrett and Dr. Wolf have shown that a much better strategy is to mix varieties and treatments, pushing the fungus in several directions at once. Simply planting a mixture of three different varieties gives an increase in yield of 8-12 percent over a uniform field. Rotating three varieties from a pool of four is better still. Treating one of the three planted with a fungicide, a different one each year, keeps the farmer one step ahead of the disease.

Facing the Issue of Pesticide Residues:

How Can Consumers Bring About Change?

BY CAROLE SUGARMAN

THE WASHINGTON POST, JUNE 3, 1987, P. E1.

Ralph Nader wrote to several large supermarket chains last year asking that they stop selling apples treated with daminozide, a farm chemical shown to cause cancer in laboratory animals. Four grocery chains, plus several apple juice and baby food companies agreed, and turned to suppliers who grew apples without daminozide. Meanwhile, the chemical is still on the market.

Thus, in the case of Alar, the trade name for daminozide, organized consumer pressure brought about stricter accountability than the government has yet required. It showed how in some cases the marketplace can move faster than the bureaucracy.

A study published two weeks ago by the National Academy of Sciences, which reported that current government regulations should be significantly improved to better protect the food supply from carcinogenic pesticides, raises the debate of what consumers can do – both personally and collectively – to bring about change and reduce their own risks.

The report picked 15 foods estimated to have the greatest potential risk. It did not imply that these 15 foods are actually dangerous as they are currently grown and marketed, but that regulations left them with the greatest allowable risk. In the case of tomatoes, for instance, not only are they widely consumed, but more carcinogenic fungicides are approved for them than for any other food. However, it is unlikely that all of those fungicides would be used at one time.

NAS used the worst-case scenario that assumed that pesticide residues were present in foods at the maximum tolerance level, that 100 percent of the acreage of the crop was treated and that exposure occurred over a 70-year lifetime.

The report "is not a food safety issue," said committee member Donald Bissing of FMC Corporation a manufacturer of agricultural machinery and chemicals; rather, the numbers were meant to identify regulatory priorities.

While the public "shouldn't stop eating tomatoes," said Richard Wiles, project officer for the NAS study, "people should be concerned." Wiles said that the fundamental problem is that the regulatory system perpetuated exposure to chemicals that are more hazardous than many safer alternatives.

There are only bits and pieces of data on actual pesticide residues in food as consumed, although more is surfacing as a result of the NAS report, according to Wiles. The most comprehensive information to date is the Food and Drug Administration's Total Diet Study, which involves the analysis of 234 foods, prepared and cooked as consumers actually eat them. In 1986, pesticide residues on these foods were well below acceptable daily intakes, according to Ellis Gunderson, chemist with the FDA's Division of Contaminants Chemistry. The study is limited, however, because it does not analyze every pesticide for which there is a regulated tolerance.

Even so, environmental groups have long criticized the methods by which tolerance levels are set. Almost half of all registered pesticides lack fundamental safety studies on which tolerances are set, a factor which leads advocates to believe that their risks are greater than currently known.

At least nine fungicides presently used are known to cause tumors in animals, according to EPA. For example, one widely used one, mancozeb, was estimated by EPA to have a risk of 2.2 cancers per 100,000 people, when it breaks down to a more dangerous chemical in the body.

From a consumer standpoint, it is virtually impossible to determine what pesticides have been used on a particular food, given the complexity of the distribution chain and fact that although numerous chemicals are often registered for a given crop, they may not actually be employed. There are striking regional differences in pesticide use, too, depending on climate and geography. The majority of fungicides used on tomatoes, for example, are applied by growers in the Midwest, East and Southeast, even though less than one-third of tomatoes are grown in these regions, according to the NAS report. Similarly, potatoes from Idaho have fewer fungicides applied to them than those from Maine.

Unfortunately, since there is very little information disclosed at the point of purchase, "protecting yourself from pesticide residues is not a self-help project," said Ellen Haas, executive director of the consumer group Public Voice for Food and Health Policy. Haas recommends that consumers be wary of "perfect-looking produce," which may have been sprayed liberally to kill insects that would mar the look of perfection and to be aware that imported produce may contain pesticides either banned in the United States or above domestic tolerance levels.

In addition, everyone agrees that washing food is likely to get rid of at least some residues. "We can't wash everything off of anything," said Wiles. "But keep washing. Wash and write your congressman."

While the EPA has very little data on what home washing actually does, Chuck Trichilo, chief of the agency's residue chemistry branch, said that warm water may be better than cold, and that mild soapy water may remove even more residues. Trichilo, who says he uses warm water and Ivory soap to wash his own produce and peels his apples because he doesn't want to eat the wax, emphasized that thoroughly washing off soap is crucial. (The June issue of the EPA Journal recommends that consumers scrub fruit and vegetables with a brush and peel them if possible; it does not recommend soap. It also suggests discarding fats and oils in broths and pan drippings, since residues of some pesticides concentrate in fat.)

When it comes to rinsing produce after it is harvested, not all crops are washed since water may decrease their lifespan, according to Claudia Fuquay, director of National Agrichemical Program of the United Fresh Fruit and Vegetable Association. Fuquay said that most head lettuces are not commercially washed; however, their outer leaves are generally removed for handling and aesthetic reasons before they are packed. Tomatoes, citrus and most tree fruits are washed, according to Fuquay.

As for washing before produce is processed, research performed by the National Food Processors Association found that 71 percent of parathion residue on spinach was moved by washing and blanching using factory sprays and equipment. NFPA concluded that the nature of the chemical and the crop on which it is applied affect the amount of pesticide that will be removed.

Unfortunately, some pesticides just won't wash off. Systemic pesticides, which lodge inside a food, cannot be removed by rinsing. Benomyl, a fungicide used on tomatoes, and daminozide in apples are two examples of systemic pesticides.

Waxes, which are often used on the surface of fruits and vegetables, frequently are mixed with fungicides, according to Sandra Marquardt, resource information coordinator with the National Coalition Against the Misuse of Pesticides. Both the wax and the fungicide are difficult to remove, Marquardt said.

Determining pesticide residues in processed food is another issue. With some foods, such as tomato paste and dried fruit, pesticide residues actually become higher after the food is processed because these products are concentrates.

According to a study from the National Food Processors Association, tomatoes containing 1.76 parts per million of benomyl after harvest were reduced to .31 ppm after commercial washing. When the tomatoes were turned into paste, the residue increased to .57 ppm.

The increase, however, was still 1/10th of the residue permitted in the raw agricultural product, according to Dennis Heldman, executive vice president of scientific affairs at NFPA.

Whatever the types and amount of pesticide residues in food and whether or not they pose a health threat, it is clear that consumers are concerned about them. According to a recent survey conducted by the Food Marketing Institute, a trade group of supermarket chains, 76 percent of shoppers rated pesticide residues in food a "serious hazards." No other food and health issue was rated as serious a hazard by as many people.

Consumer groups, who have hounded the EPA and FDA for years to better regulate pesticide residues in food, are now trying to mobilize this concern by putting the pressure on growers, manufacturers and retailers to offer pesticide-free alternatives.

Although the committee makes no recommendations for or against organic agriculture, Wiles of NAS said, "as a consumer you're helpless unless you manipulate the market. The best weapon you have is your money."

"So much time and effort goes into lobbying to get a certain chemical banned," noted Ben McKelway, associate director of the recently formed Americans for Safe Food. While the group tries to keep track of regulatory efforts, the major focus, he said, is in stimulating demand for pesticide-free foods.

ASF, coordinated by the advocacy group, Center for Science in the Public Interest, is setting up local action groups throughout the country to encourage interest in organically grown foods.

McKelway said the local consumer leader will "very politely urge supermarket buyers to seek out organic products to offer an alternative." The whole process is expected to be taken, he says, "on a positive, cordial basis."

Organic foods cost about 10 to 30 percent more than conventionally grown food, according to McKelway, but he says "once it gets rolling, we hope the price differential would decrease. Right now consumers must pay a premium for safer foods."

According to Karen Brown of the Food Marketing Institute, a number of supermarket chains have expressed a willingness to provide organically grown foods, but she noted they face several obstacles. One is the problem of availability.

Larry Johnson of Safeway said that the chain has had "some in-house discussion" about organic produce but that the biggest problem is that the quantity available doesn't match the chain's needs. Johnson did say, however, that organic produce could be promoted on a select store basis or as a specialty item.

Sue Challis, spokesperson for Giant, said the chain has investigated buying organic produce, but has found similar supply problems. Additionally, "if we had consumers calling us every day and asking for organic produce, it would [be] a different story," Challis said.

Richard Koslow, co-owner of Organic Farms, a Beltsville wholesale company that sells organic goods throughout the East Coast, said that the company has had ongoing discussions with large supermarket chains. Aside from concerns over adequate consumer demand, Koslow said, the companies are reluctant to change a "very ingrained pattern of behavior."

Another problem that supermarkets face, noted Brown, is certifying that the food was actually grown without chemicals. McKelway is well aware of the problem of certification: "There's a potential for abuse there, and I just hope that it's not too common." Several months ago, Giant Food was preparing to carry so-called "natural beef," but canceled the program at the last minute after discovering the cattle had been treated with antibiotics.

One problem is that there is no national definition of organic. McKelway hopes that differences among farmers can be ironed out enough to win support for a bill in Congress for a definition. Several states already have definitions, but they are not all the same.

Marquardt of the National Coalition Against the Misuse of Pesticides said that consumers should educate themselves about the problems of pesticides, support the organic movement and write to their representatives on both the state and federal level "to let them know that they don't want to take it anymore."

Freelance writer Patricia Picone Mitchell contributed to this report.

An Alar Apple a Day... May Be One Too Many

HRG HEALTH LETTER, JULY-AUGUST, 1986.

According to the EPA, daminozide is used on many different varieties of apples. For the major varieties, the Agency estimates that the following percentages of the fresh crop are treated with daminozide:

APPLE VARIETY	%FRESH CROP
Stayman	56%
McIntosh	55%
Golden Delicious	43%
Red Delicious	39%
Jonathan	19%
Rome	1%
All Others	less than 1%

Daminozide (also known under the Uniroyal brand name Alar) is a dangerous pesticide with highly questionable benefits. Both daminozide and its breakdown product, unsymmetrical dimethylhydrazine (UDMH), have been shown to cause cancer in animals. Therefore, most scientists agree that they are likely to cause cancer in humans as well. Furthermore, daminozide is a "systemic" pesticide. This means that it is incorporated into the fruit and cannot be peeled or washed off. The Environmental Protection Agency (EPA) has already concluded that the risks from daminozide outweigh its benefits for all food uses of daminozide. The benefits of daminozide use are largely cosmetic. Daminozide is used to delay fruit-ripening, which permits harvesting at one time and results in increased storage life, increased firmness, and heightened red color in apples. On peanut crops, daminozide produces shorter, more erect vines.

Approximately 75% of the estimated 825,000 pounds of daminozide used annually is applied to apples and 12% is used on peanuts.

In addition, children may be the most highly exposed group, since they consume larger quantities of certain daminozide-contaminated foods, such as apple juice and apple sauce, than adults. All children under the age of 12 and infants less than one year old were estimated by the EPA to be exposed to two and seven times as much daminozide, respectively, as the total population average.

Aside from widespread consumer exposure to daminozide, many agricultural workers are also endangered by the use of this pesticide on the job.

In the absence of any action by the EPA to ban or

label daminozide-containing foods, consumers have no way of knowing how to avoid this dangerous exposure. . . . [C]onsumer pressure has managed to convince some manufacturers of apple products to reject daminozide-treated apples. However, it is unfortunate that the EPA – one of the federal regulatory agencies charged with protecting the public health – has decided to turn its back on that responsibility in favor of corporate interests.

ALAR

Safer apples are on the way: an action update

HRG HEALTH LETTER, SEPTEMBER/OCTOBER 1986.

In the last issue of *Health Letter*, we told you about the dangerous "fruit cosmetic," called ALAR, Uniroyal's brand name for the apple-reddening but carcinogenic chemical daminozide. Apple and other fruit trees and some vegetables are treated with daminozide not long before harvest to make the ripening more uniform. The fact that at the peak use of this chemical less than 40 percent of apples and a much smaller proportion of other fruits and vegetables were treated with ALAR, indicates how dispensable this carcinogen is.

The Environmental Protection Agency had begun the process of banning ALAR but, under the gun of Uniroyal, they backed off, thereby protecting Uniroyal but endangering children and others who enjoy apples, apple products, and other fruits and vegetables. Because we believe that EPA is breaking the law by allowing ALAR to stay on the market, despite the fact that it is a known cancer-causing substance, a petition to ban the chemical was filed on July 2, 1986 against EPA. In addition to Public Citizen and the Natural Resources Defense Council, other petitioners included:
• the State of New York,
• the Maine Department of Human Services,
• Noah Beyeler, Naomi Beyeler and Benjamin Pourdon, three children age one year, four years, and nineteen months, living in Oakland, California who regularly consume apples, apple products and other foods that might contain ALAR and its also-carcinogenic contaminant/breakdown product, UDMH,
• Philip Landrigan, M.D., Chairman of the Committee on Environmental Hazards, American Academy of Pediatrics,
• Alan J. Finley, M.D., a pediatrician from Martinez, California, representing his patients.

As we go to press, the time given by the petitioners to EPA to respond to the petition is running out, and it is likely that it will be necessary to file a lawsuit against the agency in order to force ALAR off the market.

Major Supermarket Chains Stop Buying Alar-Treated Apples

In addition to filing the petition, we had urged you to demand of your supermarket that they not stock apples which had been treated with ALAR. Many supermarkets report many calls from consumers asking about this. Among those who called supermarkets was Ralph Nader, founder of Public Citizen and co-founder of the Health Research Group. By mid-July, Safeway, the nation's largest supermarket chain, had announced that "we do not want to buy any ALAR-treated apples." During the next several weeks, following Safeway's lead, Kroger's, the second-largest supermarket chain; Giant Foods, a Washington, D.C. chain; Grand Union; A&P; and the Detroit-area Farmer Jack's also announced that they would not purchase ALAR-treated apples.

Michigan and Washington State Apple Growers Stop Using Alar

The state apple commissions in Michigan and Washington state (trade associations of apple growers) announced that they had advised apple growers in those states not to use ALAR on their apple crops. We have been told that many other state apple-growing associations are considering a similar action.

Uniroyal Toughs It Out; Alar is "Safe and Has No Risk"

Despite all of the scientific evidence that ALAR and its breakdown product UDMH are both carcinogens, despite all of the support for banning the chemical, despite actions by apple growers, apple-purchasing supermarkets, apple juice makers (Tree Top and Motts), despite everything moving in the other direction, Uniroyal still wants to believe that its favorite fruit and vegetable cosmetic is OK. A spokesman for Uniroyal recently told a reporter that "the best scientific minds in the country have looked at it and said there's no risk. The product is safe.

Should chemicals which cause cancer in animals really worry us?

To put Uniroyal's viewpoint another way, just because large amounts of ALAR, when given to animals, cause cancer, it does not mean that humans, eating much smaller amounts, will be affected. In order to review the topic of cancer in animals, let us ask a few questions.

1. Won't most chemicals, if you feed large enough amounts of them to animals, cause cancer?

Fortunately not. Among the more than 10,000 chemicals which have been given to animals in large-dose experiments, there are fewer than 400 which have been clearly shown to cause cancer. In most of the cases where it was possible to do proper studies to see whether the chemicals that caused cancer in animals also caused cancer in humans, the human studies confirmed this.

2. But even if large doses of certain chemicals cause cancer in animals, why can't smaller doses be tested in order to more closely approximate human intake?

There are three main reasons why large doses must be used in animal experiments.

• First, animals such as the laboratory mice or rats which are typically used in such experiments, do not live as long as humans. Instead of eating apples for 60 or 80 years and getting that much of a dose of whatever chemical the apple is laced with, the animal only gets the chemical for two years. So, in order to simulate a lifetime of lower level human intake, on this basis alone, there is a clearly evident reason for large animal doses for what is a relatively short time compared to the multiple decades of human exposure.

• Second, for ethical reasons, if the questionable substance really causes cancer in humans, it is obviously not right to wait for the decades it often takes for cancer-causing chemicals to produce cancer in humans after the time exposure begins. Especially if humans are being exposed in the interim, the dose in an animal experiment should be as high as possible so as to shorten the time until the question about the chemical's carcinogenicity is answered. This latter point is related to the fact that higher doses of known carcinogens take a shorter time to produce cancer than lower doses.

• Third, high doses are used so that the problem of probability does not defeat the purpose of the experiment. If we assume that humans are equally sensitive to the possible cancer-causing properties of a chemical as are animals, and that we would be worried if even one out of a million people eating the average human intake got cancer, how would we design an animal experiment to test out this possibility? For the more than 100 million people who eat considerable amounts of apples or apple products, even one case of cancer out of a million exposed would indeed be worrisome for this would amount to approximately 100 cases of cancer. But if we are to feed the rats exactly the same dose as humans take in, unless we test millions of rats, we will not get a positive result, even on chemicals we may have other reason to know causes cancer. In fact, the typical animal experiment usually employs 50 to 100 animals in the chemical test group, rather than millions. So even if the chemical actually caused cancer in one out of a million animals or humans at the low dose, the experiment using the low dose on only 100 animals would falsely appear to be negative. To avoid this falsely negative result which could have devastating human consequences, higher doses are used on these smaller numbers of animals so that if the chemical really is a carcinogen, the experiment will be positive.

[P.S. After the EPA denied the Petition by Public Citizen and the Natural Resources Defense Council, Ralph Nader and Public Citizen, the Natural Resources Defense Council, the State of New York, Mark, Noah and Naomi Beyeler and Dr. Philip J. Landrigan filed suit on March 9, 1987 in the U.S. Court of Appeals, 9th Circuit Court, to require EPA to ban the use of ALAR.]

What can you do?

1. Urge your supermarket or grocery or fruit peddler to join the growing list of fruit and vegetable sellers who do not buy apples which have been treated with ALAR. Especially if there is a glut of apples this year, help to assure that those apples which growers cannot sell (and have to be tossed out) are the ones which have been treated with ALAR.

2. If you live in an apple-growing state with a state apple commission, urge the commission to follow the lead of Michigan and Washington and advise growers not to use ALAR anymore.

3. Urge Safeway, Kroger's and all other stores to refuse to buy any produce which has been treated with ALAR. According to EPA, 11-12 percent of peanuts, 10 percent of tomatoes and 2-4 percent of cherries are treated with ALAR.

4. When fruits and vegetables which have not been treated with ALAR reach the shelves of your local supermarket or grocery, urge the manager to put up signs indicating "Not treated with ALAR."

More Than You Ever Thought You Would Know About Food Additives...Part III

BY PHYLISS LEHMANN
FDA CONSUMER, JUNE 1979

"Not the same thing a bit!" said the Hatter. "Why you might as well say that 'I see what I eat' is the same as 'I eat what I see'!"

– From Alice in Wonderland

"In sight, it must be right."

– Advertising slogan of a Midwest-based fast food chain.

The Mad Hatter lived more than a hundred years ago in the imagination of Lewis Carroll. Yet his seemingly nonsensical remark about eating what one sees and seeing what one eats makes a lot of sense today. That's because today food additives are used to make foods look, taste, and even feel the way we want them to look, taste, and feel.

Some of the additives are put into foods simply to make them more appealing. Others are there to aid in the processing and preparation of the foods. These additions rouse more than a little suspicion among eaters. That's why the fast food chain uses the slogan about preparing its food out in the open so every one can see it.

This article, the third in a series on food additives, will take a look at those two groups of additives: substances used to spark the color or taste of foods, and those that make foods behave the way we expect them to even after they've left the manufacturing plant.

Making food more appealing

Four classes of additives are used to heighten the appeal of foods: colors, flavors, flavor enhancers, and sweeteners.

Colors

Coloring agents add controversy along with color. They add controversy because they are used solely to improve appearance. They contribute nothing to nutrition, taste, safety, or ease of processing. And some consumer advocates argue that food is often made to look more appetizing at the risk of increasing health hazards.

Today food colors are used in virtually all processed foods. While their use is not restricted, per se,

they cannot be used in unnecessary amounts or to cover up unwholesome products. Artificial colors must be listed as ingredients in all foods except butter, ice cream, and cheese. There are 35 colors currently permitted for use in food. Nearly half of them are synthetic colors, which are created in laboratories. The man-made colors find the widest use because they are stronger than natural colors and thus can be used by manufacturers in smaller quantities and at less cost.

The first food colors were generally harmless vegetable dyes. In the 19th century days of the Mad Hatter, toxic mineral pigments, such as lead and copper, were used to change the color of foods.

Testing and certification of food colors were first required in the original Food and Drugs Act of 1906. Today food colors are regulated under 1960 legislation that makes the food industry responsible for proving the safety of the additives.

Synthetic food colorings usually refer to the coal-tar dyes, which are laboratory creations that are chemically different from anything found in nature. (They are now derived from petroleum rather than from coal tar.) For identification, the colors are assigned initials, the shade, and a number. FD&C Red No. 40, for example, indicates that it is a red coloring used in food, drugs, and cosmetics.

In the past half dozen years, FDA has prohibited four colors from use in foods: A violet used to stamp meats; Red No. 2, which was suspected as a carcinogen; Red No. 4, used in maraschino cherries and shown to cause bladder lesions and damage to adrenal glands in animals; and Carbon Black, used in candies such as licorice and jelly beans.

FDA also proposed to prohibit Orange B, used in sausage and hotdog casings, because of possible contamination with a carcinogen. The manufacturer voluntarily stopped producing it in 1978.

The two most widely used food colors now are Red No. 40 and Yellow No. 5, and both are under fire because of reports of possible health risks. Red No. 40 is suspected of causing premature malignant lymph tumors when fed in large amounts to mice. In 1977, the Health Research Group petitioned FDA to prohibit its use along with several other colors. FDA denied the petition, but tabled a final decision on Red No. 40 pending a study review.

The problem with Yellow No. 5 is that it causes allergic reactions – mainly rashes and sniffles – in an estimated 50,000 to 90,000 Americans. The reactions are usually minor but in some instances can be life threatening. Because of its relatively narrow effect, FDA has proposed a regulation to require manufacturers to list Yellow No. 5 on the labels of any food products containing it.

Flavors

Some 1,700 natural and synthetic substances are used to flavor foods, making flavors the largest single category of food additives. Most of the flavors are synthetic because they are cheaper than the real McCoy and because there probably would not be enough of the real McCoys – strawberries, for example – to produce all the strawberry flavoring desired by diners. Artificial flavorings are usually not derived from a single chemical but are the result of a complex process that involves analyzing the individual chemicals present in a flavor, reproducing those in a laboratory, and synthesizing them to create a taste approximating the real thing.

Flavors are listed on food labels in general terms, such as "artificial flavor" or "spices." If a product contains any added flavoring, either natural or synthetic, that fact must be noted on the label. For example, a label that says "strawberry yogurt" means that the product contains all natural strawberry flavor. "Strawberry-flavored yogurt" indicates that it contains natural strawberry flavor plus other natural flavorings. "Artificially flavored strawberry yogurt" means that it contains only artificial flavorings or a combination of artificial and natural flavors.

Flavors have come under less criticism than colors, perhaps because they serve a more direct purpose in foods. Still, some consumer groups question the necessity of using artificial flavors. FDA scientists maintain that anyone sensitive to artificial flavors would be likely to react to natural ones as well because of the chemical similarities.

A few flavorings have been prohibited in the past 10 to 15 years because of health hazards. Probably the best known outcast was safrole, the principal flavoring in sassafras root, once widely used in root beer. FDA banned safrole after tests showed it caused liver cancer in rats. Coumarin, used as an anticoagulant drug, was once present in imitation vanilla extract and other flavorings but was banned for food use because large amounts could cause hemorrhaging.

Flavor Enhancers

These compounds magnify or modify the flavor of foods and yet do not contribute any flavor of their own.

Some of them work by temporarily deadening certain nerves – those responsible for perception of bitterness, for example – thereby increasing the perception of other tastes.

The best known flavor enhancer is the amino acid, monosodium glutamate (MSG), widely used in restaurants and in prepared foods. Scientists are not sure exactly how it works, but suspect that it increases the nerve impulses responsible for perception of flavors. Several years ago, public pressure persuaded manufacturers to stop using MSG in baby foods after studies showed that large amounts had destroyed brain cells in young mice.

MSG also produces the so-called "Chinese restaurant syndrome," which causes some people to have a burning sensation in the neck and forearms, tightness in the chest, and headache after they consume the relatively large amounts of MSG often used in food served in Chinese-style restaurants.

Sweeteners

Though technically flavors, sweeteners are generally considered a separate category. They are among the most commonly known food additives. Who has not heard of saccharin?

Sweeteners are classified as nutritive and non-nutritive. The nutritive ones, metabolized by the body to produce energy, include the natural sugars such as sucrose (common table sugar), glucose, and fructose, as well as sugar alcohols, such as sorbitol and mannitol. Non-nutritive sweeteners, which are not metabolized and therefore contribute no calories to the diet, include cyclamate, which is currently prohibited from use in food, and saccharin.

Natural sugars are widely used in foods, not just as sweeteners but also to create a heavier mouth feel in soft drinks and as browning agents in baked goods. Some consumer groups oppose adding sugar to food because they say it represents "empty" calories devoid of vitamins, minerals, or protein. They also argue that sugar contributes to tooth decay.

The sugar alcohols, chemical variants of natural sugars, have been around for decades but have been promoted in recent years as "low-cal" alternatives to sugar and as less likely to cause tooth decay. Foods containing these sweeteners, are not truly low-calorie. Nor can they be considered completely "free" foods for diabetics, as they can and do lead to the production of some blood sugar. An FDA regulation scheduled to become effective in 1980 will require manufacturers to state on labels that inclusion of these nonsugar sweeteners does not mean the product is "low-cal" or "reduced calorie."

The most widely used sugar alcohol is sorbitol, put in chewing gum, mints, candies, and dietetic ice cream. Though safe, it does have a laxative effect in large amounts. Mannitol, on the other hand, may cause diarrhea in relatively small amounts because it accumulates water during its extremely slow passage through the intestinal tract. This side effect has caused mannitol's use to be limited to the powdery coatings on some chewing gums.

Xylitol, another sugar alcohol promoted several years ago for sugarless gum and dietetic food, fell from favor following several negative health reports and studies. Some manufacturers have voluntarily stopped using it.

The non-nutritive sweeteners – cyclamate and saccharin – have proved to have a taste for controversy. FDA banned cyclamate in 1969 after studies indicated it caused cancer in animals. The manufacturer has sought reinstatement, citing new studies, but so far it remains off the store shelves.

Saccharin was originally on the GRAS (Generally Recognized As Safe) List but was removed in the early 1970s when evidence of health hazards began to mount. In April 1977, FDA proposed to ban saccharin as an additive in food (it's primarily used in diet sodas) following a Canadian government study that showed it caused bladder tumors in rats. The ban would have allowed the continued use of saccharin as a tabletop sweetener if the industry could prove it was beneficial to diabetics or to people on weight-reduction diets.

The public outcry was such you would have thought FDA was trying to take real candy from all the nation's babies. Congress heard the outcry and passed an 18-month moratorium on the FDA ban proposal. The moratorium ended May 23, 1979, and FDA has announced its intention to re-propose the ban – a regulation-making process that could take 15 to 18 more months.

Preparing and Processing Foods

The final category of food additives consists of those used in the preparation and processing of foods. These additives are used by manufacturers to get desired effects during processing and beyond. To the consumer, the additives give the food some of the characteristics that are associated with the products.

The function of these additives are many. Some cause baked goods to rise. Others prevent ice crystals from forming in ice cream and keep peanut butter from separating into oily and dry layers. Because of such additives, shredded coconut stays fresh and moist in the can.

Of the four major categories of food additives, these are the least clouded by controversy. Consumer groups caution about using some specific compounds, but there is nothing like the furor generated by other additives such as saccharin and nitrites.

There are seven major groups of additives that are considered aids in processing or preparation of foods:

Emulsifiers (Mixers)

Some liquids don't mix unless there is an emulsifier around. In salad dressing, for example, oil and vinegar normally separate as soon as mixing stops. When an emulsifier is added, the ingredients stay mixed longer. In pickles, beverages, and candies, emulsifiers help disperse flavors and oils that otherwise would not be soluble in water. Without these compounds, ice cream and other frozen desserts would separate and lose their creamy texture. In baking, emulsifiers improve the volume and uniformity of breads and rolls as well as make batter and dough easier to handle.

Many emulsifiers come from natural sources. Lecithin, naturally present in milk, keeps fat and water together. Egg yolks, which also contain lecithin, improve the texture of ice cream and mayonnaise. The mono- and diglycerides come from vegetables or animal tallow and make bread soft, improve the stability of margarine, and prevent the oil and peanuts in peanut butter from separating.

Several emulsifiers have elicited concern about health risks. The Center for Science in the Public Interest, in its poster "Chemical Cuisine," advises consumers to avoid brominated vegetable oil (BVO), used to maintain the characteristic cloudy appearance of citrus-flavored drinks by keeping oils in suspension. The Center warns that residues from the additive accumulate in body fat. Because of this problem, FDA has removed BVO from the GRAS List and set specific levels at which it may be used.

Stabilizers and Thickeners

These compounds "improve" the appearance of food and the way it feels in the mouth by producing a uniform texture. They work by absorbing water. Without stabilizers and thickeners, ice crystals would form in ice cream and other frozen desserts and particles of chocolate would settle out of chocolate milk.

Stabilizers also are used to prevent evaporation and deterioration of the volatile flavor oils used in cakes, puddings, and gelatin mixes.

Most stabilizers and thickeners are natural carbohydrates. Gelatin – made from animal bones, hooves, and other parts – and pectin – from citrus rind – are used in home and commercial food processing. Extra pectin, for example, is added to thicken jams and jellies.

In the past 30 years, vegetable gums – from trees, seaweed, and other plants – have become widely used as thickeners. These are so effective that 0.1 percent can produce the same degree of thickness in water as a high concentration of starch. One problem is that some, such as tragacanth gum and gum arabic, cause allergic reactions in a few susceptible people. The

Center for Science in the Public Interest warns consumers especially about tragacanth gum, which it says has caused some sever allergies. FDA does not believe this problem is any more prevalent than allergies from eggs, chocolate, milk, or other natural foods.

pH Control Agents

These affect the texture, taste, and safety of foods by controlling acidity or alkalinity. Acids, for example, give a tart taste to such foods as soft drinks, sherbets, and cheese spreads. A more important use is to insure the safety of low-acid canned foods, such as beets. Normally, these low-acid foods have to be cooled longer at higher heat than acidic foods to render them sterile because they are more receptive to the bacteria that cause botulism. By adding acids, manufacturers can eliminate the need for extra heat that might detract from the marketable quality of the food. Natural organic acids, such as citric, fumaric, tartaric, and malic acids, are used in canned foods, although mineral acids, such as hydrochloric, are preferred in some cases.

Alkalizers alter the texture and flavor of many foods, including chocolate. After cocoa beans are picked, they are allowed to dry and ferment before they are made into chocolate. During processing, alkalizers are added to neutralize the acids produced during fermentation and to provide a darker, richer color and milder flavor in the finished product.

Leavening Agents

Although air and steam help create a light texture in bread and cake, carbon dioxide is the key to making baked goods rise properly. Without leavening agents that produce or stimulate production of carbon dioxide, we would not have light, soft baked goods.

The earliest leavening agent was yeast, which produces carbon dioxide through fermentation. Today two other leavening agents are found in home kitchens and commercial bakeries. One, baking soda (sodium bicarbonate), releases carbon dioxide when heated. Usually an acid ingredient, such as sour milk, is used along with baking soda to eliminate the soapy-tasting byproduct of this chemical reaction. The second is baking powder, a combination of sodium bicarbonate and acid salts that react in the presence of water to produce carbon dioxide.

Maturing and Bleaching Agents

Maturing and bleaching agents are used primarily to get flour ready for baking because natural pigments give freshly milled flour a yellowish color. Flour also lacks the qualities necessary to make a stable, elastic dough. When aged for several months, it gradually whitens and matures to become useful for baking.

In the early 1900's scientists discovered they could hasten bleaching and maturing – and eliminate costly storage – by adding certain chemicals. These agents do not remove anything from the flour and leave little residue. They simply change the yellow pigments to white and develop the gluten characteristics necessary for baking.

Bleaching agents, such as benzoyl peroxide, also are used to whiten milk used for certain cheeses known for their whitish curd, such as blue cheese and gorgonzola. Bleaching is considered necessary because the grass that cows eat causes them to yield buff-colored milk.

Anti-caking Agents

Compounds such as calcium silicate, iron ammonium citrate, and silicon dioxide are used to keep table salt, baking powder, confectioner's sugar, and other powdered food ingredients free flowing. By absorbing moisture, these chemicals prevent caking, lumping, and clustering that would make powdered or crystalline products inconvenient to use.

Humectants

Humectants are substances that retain moisture in shredded coconut, marshmallows, soft candies, and other confections. One of the most common is glycerine. The sweetener sorbitol also is used for this purpose.

Although these are the major additives used in processing and preparation, there are other additives with specialized uses, such as clarifying agents, which remove small mineral particles that cloud such liquids as vinegar; firming agents that help coagulate certain cheeses and improve the texture of pickles, maraschino cherries, canned peas, tomatoes, potatoes, and apples; foam inhibitors that prevent foam formation on pineapple juice or on other foods during washing, cooking, or processing; and sequestrants that chemically "hold" minerals in soft drinks that might otherwise settle out and cloud the beverage.

Phyllis Lehmann is a freelance writer.

Dyes in Your Food

The Government has Refused to Take Dangerous Dyes Out of Your Food, Drugs and Cosmetics, But There Are Several Steps You Can Take to Protect Yourself and Your Family

HRG HEALTH LETTER, MARCH/APRIL 1985.

What do Strawberry Jello, Orange Koolaid, raspberry soda, grape popsicles, much candy and baked goods, most brands of ice cream, maraschino cherries, many snack foods, and most pet food have in common? They are laced with millions of pounds of artificial coal-tar based dyes with names such as Red 3, Blue 1, Blue 2, Green 3, Yellow 5, Yellow 6, and Red 40. Each year, Americans eat (as food dyes), swallow (as pill coatings or medicinal syrups) or rub on themselves (as cosmetics) 6.4 million pounds of these 7 dyes, mostly in food. Four of theses food dyes (Red 3, Yellow 5, Yellow 6 and Blue 2) which make up over half of the food dyes used each year have been shown to cause cancer as have other dyes which are not used in food but are used in drugs or cosmetics.

Table 1: Disappearing food dyes:

17 GONE – MORE TO GO

17 U.S. Food dyes now banned, delisted or not produced

DYE	YEAR BANNED, DELISTED OR LAST PRODUCED*
Sudan 1	1918
Butter Yellow	1918
Red 32	1956
Orange 1	1956
Orange 2	1956
Yellow 1	1959
Yellow 2	1959
Yellow 3	1959
Yellow 4	1959
Red 1	1961
Green 1	1966
Green 2	1966
Violet 1	1973
Red 2	1976
Red 4	1976
Citrus Red 2	1976*
Orange B	1978*

Children and food dyes

In July 1976, the FDA Division of Consumer Studies published a report concerning the ingestion of food dyes by children in the United States. Some of their findings follow:

• 95-99% of children eat some food containing coal-tar (petroleum-derived) food dyes.

• Over four million children will have consumed a total of more than one pound of coal-tar food dyes by the time they are 12 years old.

• The maximum consumption of food dyes by children is as high as three pounds by age 12.

These figures are quite conservative since there has been a 50% increase in the total use of food dyes in the past ten years. In addition to the possibility that they are more susceptible to carcinogenic chemicals such as food dyes, children will have a longer history of ingestion and thus a greater likelihood of developing cancer because they eat these dyes.

History of food dyes

Although we are very concerned about any carcinogenic or otherwise toxic dye, food dyes are the biggest worry because much larger quantities of these dyes are produced and ingested than drug or cosmetic dyes and because children are major targets/victims of food dyes.

The history of food dyes in the United States is really a history of disappearing food dyes. Of the 24 food dyes which at one time have been allowed in the American food supply, 17 are now banned, delisted or no longer produced as can be seen in Table 1.

Of the seven food dyes still used (see Table 2), three (Red 3, Yellow 5 and Yellow 6) are the subject of a Public Citizen Health Research Group petition to FDA for a ban and a subsequent law suit. These three alone, however, accounted for 3,396,855 pounds or 53% of all food dyes certified in fiscal 1984. The other four dyes also have serious questions about their safety as indicated in Table 2. Of these four, Blue 2 is currently the subject of an FDA Administrative Hearing as a result of Public Citizen's objections to the FDA's decision to permanently list it. There was a statistically significant increase in the number of brain tumors in animals fed Blue 2. We are currently reviewing the safety studies on the other three dyes (Green 3, Red 40 and Blue 1) because of the toxic findings listed in Table 2.

It is likely that in a very short time, there will be few, if any "certified" petroleum-derived food dyes left. Fortunately, products such as Dannon Yogurt ("No artificial anything"), General Foods Cheerios (no Yellow 5 as in the past), and many other foods without artificial colors have shown, by their extraordinary success, that we can all do very well without these dyes.

What has Public Citizen done?

In December, 1984 Public Citizen Health Research Group petitioned FDA to ban the use of ten widely-used food, drug or cosmetic dyes. On January 22, 1985 we sued the government to remove these dyes, all of which have been shown to cause cancer, from the market. None of these dyes has ever been found safe by FDA for all uses but all have remained on the market pending resolution of their "provisionally listed" status. For six of the ten dyes, FDA recommended such a ban. In March, 1984, we argued that as long as these dyes remain on the market, FDA is violating the safety requirements of the Color Additive Amendments to the Food, Drug and Cosmetic Act. By failing to immediately ban these dyes, the Reagan Administration is making a mockery out of its alleged cancer reduction goals and is completely demoralizing dozens of FDA employees who know these dyes are too dangerous for continued use by the public.

We obtained a copy of an internal FDA memo confirming the illegality of marketing five of these dyes (D&C Reds 8, 9, 19, and 37 and D&C Orange 17) written by FDA's Director of the Center for Food Safety and Applied Nutrition Dr. Sanford Miller on November 23, 1984. Dr. Miller, referring to the cosmetic industry's plea for further peer review of the dyes, said:

In other words, the only effect that we can realistically see from additional peer review is a further delay that would risk a lawsuit asserting that FDA is not adhering to its responsibilities under the law. In our judgment we have already extended the provisional list so many times for such tenuous reasons that we are in danger of losing both a lawsuit and our credibility as a regulatory agency.

What can you do?

1. Avoid using any foods with artificial dyes ("artificial coloring," "U.S. Certified colors") such as the seven listed in Table 2: Red 3, Blue 1, Blue 2, Green

Table 2: 7 Remaining Food Dyes in U.S.

DYE	MAIN USES	POUNDS CERTIFIED 1984	TOXICITY FINDINGS	COMMENTS
RED 3	candy, desserts, baked goods	241,265	thyroid tumors, chromosomol damage	FDA recommended ban
RED 40	beverages, candy, desserts, pet food	2,630,578	earlier lymphomas (lymph tumors)	banned in EEC
BLUE 1	beverage, candy, baked goods	260,417	chromosomol damage	banned in France, Finland
BLUE 2	pet foods, candy, beverages	101,223	brain tumors	banned in Norway (pending FDA hearing)
GREEN 3	beverages, candy	3,597	bladder tumors	banned in EEC
YELLOW 5	pet food, beverages, baked goods	1,620,540	allergies, thyroid tumors, lyphocytic lymphomas, chormosomal damage	banned in Norway
YELLOW 6	beverages, candy, desserts, sausage	1,530,050	allergies, kidney tumors, chromosomal damage	banned in Norway, Sweden

TOTAL 6,392,670

3, Yellow 5, Yellow 6 and Red 40. More and more nationally distributed foods such as Cheerios, Grape-Nuts, Nutri-Grain, Dannon Yogurt, Pepperidge Farm Products and others not only contain no artificial coal tar food dyes, but also have no other additives. Buy these instead of dye-containing products. In addition, real fruit juices and gelatin desserts or popsicles made with fruit juice are healthy substitutes.

2. The seven drug and cosmetic dyes which should be avoided, all "D&C" (drug and cosmetic dyes) are: Red 8, Red 9, Red 19, Red 36, Red 37 and Orange 17. Read the ingredient list on cosmetics (if it is there) before using them.

3. Avoid yielding to arguments that only one in a million people using each of these dyes will get cancer at the doses used in foods, drugs, or cosmetics. For all of these dyes, for all people using them, this amounts to well over 1000 cases of cancer. In addition, hundreds of thousands of people are extremely allergic to Yellow 5 or Yellow 6 and small amounts, even a few thousandths of a gram, can cause severe reactions in some of these people.

4. Write your Senator and Congressperson asking them what they are doing to push the government to obey the food safety laws and ban these dyes.

Consumer Group Discloses Levels of Cancer-Causing Chemical in Alcoholic Beverages

CENTER FOR SCIENCE IN THE PUBLIC INTEREST
PRESS RELEASE - DECEMBER 12, 1987

Tainted Booze, a 70-page report published today by a non-profit health advocacy organization, discloses the amounts of *urethane*, a potent cancer-causing contaminant, in over 1,000 alcoholic beverages.

The report, released at a Washington press conference by the Center for Science in the Public Interest (CSPI), is the only guide available for consumers purchasing alcoholic beverages for holiday gifts and entertaining.

According to tests conducted by government and industry, more than 100 wines and liquors sold in the U.S. contain urethane at levels that were declared illegal over two years ago by the Canadian government. The U.S. Food and Drug Administration (FDA) has not removed any beverages from the marketplace or even publicized their names.

The consumer group also sent a letter to FDA commissioner Frank Young in which it cited statements by FDA scientists who maintain that the Canadian estimates of risks from urethane are 10 to 100 times too low.

Michael Jacobson, executive director of CSPI and co-author of *Tainted Booze*, said, "People should be able to drink without imbibing powerful carcinogens. The FDA is derelict in delaying, seemingly endlessly, the protection of consumers."

The most heavily and frequently contaminated products are bourbons, sherries, and fruit brandies. Contaminated beverages include certain Gallo and Christian Brothers sherries; Almaden and Paul Masson table wines; Jim Beam, Old Taylor, and Early Times bourbons; and Foster beer.

In addition to the beverages that exceed the Canadian limits, the guidebook lists hundreds of other whiskeys, table and dessert wines, brandies, and liqueurs that contain measurable and potentially hazardous amounts of this carcinogen. Only a few beers are contaminated.

Urethane develops accidentally during production; it is not an additive. It occurs in both domestic and imported products.

Marvin Schneiderman, a cancer specialist with the Board on Environmental Sciences and Toxicology of the National Academy of Sciences-National Research Council, said, "There is no doubt whatsoever that urethane poses a real cancer risk to regular drinkers. My concern is whether regulatory action will be taken promptly enough to save some lives."

The FDA's top food toxicologist has labeled urethane "first, second, or third" among substances that pose cancer threats to humans.

In November, 1986, CSPI petitioned the FDA to (a) systematically test products for urethane levels and inform the public of the results; (b) set strict limits on the urethane content of alcoholic beverages; and (c) recall highly tainted products, including those in storage.

Tainted Booze includes urethane levels for 1,200 government- and industry-tested samples. CSPI obtained the data under the Freedom of Information Act. Some 200 products listed contain no detectable urethane.

The report also offers CSPI's answers to the questions most frequently asked by reporters and concerned consumers.

"Until the FDA protects consumers from the threat of urethane-tainted wine and liquor, consumers need basic information so they can try to protect themselves," explained co-author Charles Mitchell, a CSPI attorney.

Tainted Booze is available for $3.95 from CSPI-Tainted Booze, 1501 16th Street, N.W., Washington, D.C. 20036.

CSPI, an 80,000 member organization based in Washington, has led national efforts to reform policies concerning alcoholic beverage ingredients, labeling, advertising, and taxation.

EXCERPTS FROM

The Goldbecks' Guide to Good Food

BY NIKKI AND DAVID GOLDBECK
(NAL BOOKS: NEW YORK, 1987)

A Review of Traditional Packaging Materials

No packaging, it seems, is perfect. While glass is the safest in terms of chemical migration, it is heavy and breakable.

In individual tests done on canned foods, research has shown that tin and iron show up in the contents. This is considerably increased when foods are placed in the refrigerator in opened cans. The problem can be minimized by applying an enamel or lacquer coating to the inside of the can to prevent corrosion.

Lead used to solder cans is another issue that has received widespread publicity. The toxic effects of this metal have long been recognized. Lead can cause anemia and it can damage the kidneys and the central nervous system in children, and the peripheral nervous system in adults. Public awareness about the potential dangers has put pressure on the food industry to produce seamless two–piece cans or cans with electrically welded side seams which do not require lead solder, and by the end of the twentieth century, if not sooner, lead–soldered cans will probably be extinct.

The switch from steel to aluminum cans that has been embraced by the beer and soft drink sector introduces a new concern. Aluminum is another toxic metal and we have found very little data on its transfer from container to contents.

While paper seems to be an innocuous material, to present an honest picture we must inform you that several of the chemicals used in the production of paper packaging are regulated because of known hazards they can impose on health. Included in this category are polychlorinated biphenyls (PCBs), several defoaming agents which contain a formaldehyde compound, and "slimicides" used to control slime on the paper surface. Furthermore, there are a multitude of other processing adjuncts which are used in or on the paper, on coating materials, and in the glues and adhesives used to construct packages. The list of possible packaging additives in the Code of Federal Regulations is so extensive you begin to wonder how many of these could possibly have been tested during the short time packaged foods have been sold in their current volume.

Plastics

Of all packaging materials currently in use, plastics present the most concern in terms of human exposure to carcinogens. Whether rigid or soft, they contain "loose" chemicals that can leach into foods. Foods with a high fat content appear to be most susceptible and the application of heat, as in boil– and bake–in bags, also seems to initiate higher levels of chemical migration. According to an official at the FDA, the agency "works from the assumption that something from the plastic will always leach into the food."

Certainly the history of plastic packaging materials has not been very reassuring. One of the earliest hazards which came to light in the 1970s was the presence of carcinogenic vinyl chloride polymers (like PVC) in many varieties of plastic from pliable film wraps, to semirigid and rigid bottles, blister packs, cap liners, and even tubing used in the transport of some foods. While restrictions have been placed on its use, vinyl chloride still remains in some, but not all, plastic wraps (including Saran and Reynold's Plastic Wrap), plastic containers used for cooking oils, and plasticized coatings for cans and paperboard. It is also still part of the barrier layer of many plastic laminates, but there it is generally separated from the food itself by another plastic which forms the actual food contact surface. Moreover, the FDA has proposed rules that would open the way for wider use of PVC in packaging food and beverages in the future.

After thirty years of use in plastic bottles, studies also revealed substantial migration of the suspected carcinogen acrylonitrile into beverages. It was banned for use in soft drink containers after considerable stalling, but retained with limits for use in many other plastic packages. Now new formulations are reinstating acrylonitrile in beverage containers.

Another common plastic component, which goes by the initials DEHP and is actually a plasticizer used to make PVC more flexible, has also been linked to cancer, but as of 1987 no action has been taken to prohibit it. DEHP is commonly found in plastic wraps used for meat, cheese, fruits, and vegetables, as well as in bottle caps and food container liners.

A host of other chemicals, many that are prohibited as direct additives in food products, are employed in plastics. Most food processors who use these containers are probably not even aware of their presence.

Delaney Clause Victory

HRG HEALTH LETTER, DEC. 1987, PP. 10-11.

In December 1984, Public Citizen petitioned the Food and Drug Administration (FDA) to ban 10 color additives because of serious safety problems including the risk of cancer. In a major setback for the Reagan administration's cancer policy affecting 7 of the 10 dyes, the federal court of appeals in Washington, D.C. has handed consumers an important victory by ruling that the Food and Drug Administration violated the law when it approved two color additives (Orange No. 17 and Red No. 19) that cause cancer in animals. The Court unanimously held that the FDA must abide by the Delaney Clause, which prohibits the agency from approving food additives that it has determined are animal carcinogens.

The Delaney Clause is the nation's most famous public health law, and the Reagan administration targeted it for extinction shortly after President Reagan took office in January 1981. Almost immediately, the food industry, with the administration's support, proposed legislation which would have repealed the Delaney Clause. When Congress refused, the FDA adopted its de minimis policy, which loosely translated meant that the government could ignore the Delaney Clause if it concluded that the risk of cancer was very small. The Court held that this de minimis interpretation was illegal.

The decision is expected to have a broad impact on the regulation of animal carcinogens at the Food and Drug Administration. In addition to the two color additives at issue in this case, the decision is likely to result in a ban on five other dyes, including the widely-used food dye Red No. 3. It may also lead to a ban on methylene chloride, a food additive used to decaffeinate coffee. Finally, and perhaps most important, it will close the door on the FDA's ability to approve color additives and food additives in the future if those substances cause cancer in animals.

As Judge Williams noted in his opinion, Congress adopted the Delaney Clause in 1960 because it was "truly alarmed about the risks of cancer" and because it "perceived color additives as lacking any great value." While the FDA may be correct that the risk to a single individual of getting cancer from a single dye is small, the Delaney Clause recognizes that small risks add up and that as a public policy matter our society should not accept any additional cancers where the benefits of a substance are very small or nonexistent.

Undoing Delaney
FDA Allows Free Use of Dangerous Additives

BY MICHAEL F. JACOBSON, PH.D.
NUTRITION ACTION HEALTHLETTER, SEPT.-OCT. 1985, P. 7.

The Food and Drug Administration, ostensibly the protector of our food supply, is undoing the Delaney Clause, the prime legal barrier between cancer-causing food additives and the public's stomach.

In the 1950s, Congress held a series of hearings in response to the burgeoning use of synthetic food additives. The result was a 1958 law covering food additives and a 1960 law specifically for colorings. The laws required industry to prove additives safe before they could be used in foods. (Previously, industry could use additives until government proved they were unsafe). Additionally, the law gave industry two and a half years to prove that the dyes then in use were safe.

Most importantly, though, the law clarified that the public, speaking through Congress, would not tolerate food additives that increase the risk of cancer. In what has become known as the Delaney Clause (named after New York Congressman James Delaney), Congress stipulated that:

"no additive shall be deemed to be safe if it is found to induce cancer when ingested by man or animal, or if it is found, after tests which are appropriate for the evaluation of the safety of food additives, to induce cancer in man or animal..."

This put sharp teeth into the law's more general safety provisions, which themselves could be construed to ban cancer-causing substances.

Food and chemical companies perceive the Delaney clause as a constant threat, fearing that one of the additives they use may be discovered to cause cancer in animals. In fact, though, the Delaney clause actually has been invoked only a handful of times. The spirit, if not the letter, of Delaney, however, has led to the banning of several chemicals, including cyclamates and colorings, that did not strictly come under Delaney's umbrella.

Industry has fought for years to get the law changed, but each time has been stymied in Congress. For the past several years, Senator Orrin Hatch (R-Utah) has advocated legislation that would de-fang the Delaney amendment. But other legislators have blocked Hatch's efforts.

Rather than waiting, perhaps endlessly, for new legislation, industry took a back-door route. It has gotten the Reagan administration to re-interpret the existing law.

The government's handiwork is most obvious in the case of food dyes. FDA scientists and commissioners have concluded several times that Red No. 3 dye promotes cancer. But higher officials in the Department of Health and Human Services and the White House always prevented FDA from banning the dye. The government's disdain for the law was documented by a meticulous investigation and June 1985 report by the House Subcommittee on Investigations and Oversight, which is chaired by Rep. Ted Weiss (D-N.Y.)

In June, FDA granted yet another extension, of as long as six years, for Red No. 3. FDA acknowledged the connection between the dye and cancer, but said that the risk to the public was minimal. This marked the first time that FDA sanctioned the use of a cancer-causing additive, though previously it had okayed additives that were themselves safe, but that were contaminated with small amounts of carcinogens.

The government's new policy is being challenged in the courts by Public Citizen Health Research Group, but meanwhile industry is cheering.

For the public, of course, there is no reason to cheer, because the new policy means that chemicals that increase the risk of cancer (even if only slightly), will be present in our foods. We have two courses of action. First, we can let our legislators know our feelings about cancer-promoting additives. Second, we can try to avoid additives that may promote cancer.

◆

Here are some frequently used additives that may increase the risk of certain kinds of cancer:

Aspartame. A 1970s test linking the newest artificial sweetener to brain tumors in rats has never been adequately disproved.

BHA (butylated hydroxyanisole). A Japanese test found that this antioxidant causes cancer of the forestomach in male and female rats.

BHT (butylated hydroxytoluene). Depending on the experimental conditions, this antioxidant either increases or decreases tumor risks in rodents.

Blue No. 2 dye. FDA acknowledges that the evidence about the chemical's safety is equivocal, but maintains that there is sufficient evidence of safety.

Propyl gallate. The best tests on this antioxidant indicated an increased incidence of cancer.

Red No. 3 dye. FDA officials have tried to ban this artificial food coloring because it caused thyroid tumors in rats. Red No. 3 can be found in maraschino cherries, pistachio nuts, and many other foods.

Saccharin. FDA tried to ban this artificial sweetener because it causes bladder cancer.

Sodium nitrite. Used as a preservative in processed meats, this substance reacts with certain amino acids to form cancer-causing nitrosamines. Tests performed on bacon have revealed the presence of nitrosamines.

Yellow No. 6 dye. Industry has reported to FDA that this widely used dye appears to increase the number of kidney and adrenal gland tumors in rats.

Lawsuit Seeks To Prohibit Use Of Carcinogen Methylene Chloride In Decaffeinated Coffee

HRG HEALTH LETTER, JULY-AUGUST 1986, P. 13.

Public Citizen and the Consumer Federation of America have just filed suit in the United States District Court for the District of Columbia to force the Food and Drug Administration to prohibit the use of methylene chloride to decaffeinate coffee. Methylene chloride is a chemical solvent used to take the caffeine out of coffee. It is used in the majority of decaffeinated coffees currently sold in this country. Since 1967, the FDA has allowed the use of this food additive as long as the residual amount in the coffee does not exceed 10 parts per million.

Recent scientific studies conducted by the government's own National Toxicology Program (part of the Department of Health and Human Services) and by the chemical industry show that methylene chloride causes liver and lung cancer in laboratory animals, and that it may cause other kinds of cancer. On December 18, 1986 HHS and the FDA announced the "methylene chloride is an animal carcinogen." Although they proposed banning the use of methylene chloride in cosmetics such as hairsprays, they announced that they had decided to allow the continued use of the substance in decaffeinated coffee, at the same level allowed since 1967.

The lawsuit points out that a provision of the Federal Food, Drug and Cosmetic Act, known as the "Delaney Clause," prohibits the use of any food additive that has been shown to cause cancer in animals. Until June 1985, the FDA consistently took the position that the Delaney Clause required it to ban animal carcinogens from the food supply. However, applying a new interpretation of the law, in deciding to leave methylene chloride in decaffeinated coffee, the FDA adopted a "de minimis" exception to the Delaney Clause. Relying on its calculations that only a "small" number of cancers will be caused by exposure to the substance in decaffeinated coffee, the FDA concluded that it has the authority not to ban the substance.

According to internal government memos obtained by Public Citizen, a mere month before the FDA announced its decision that the currently 10 ppm level for methylene chloride was "de minimis" the agency had concluded that coffee manufacturers who use the substance can substantially reduce this residual amount. A draft memorandum dated November 15, 1985 from the FDA Commissioner to the Secretary of HHS stated that "data has been submitted to the FDA which strongly suggests that methylene chloride residues in decaffeinated coffee can be effectively lowered to a level of 0.01 parts per million using modern production processes." This would be a 1000-fold reduction from the 10 ppm level currently allowed. A second memo dated November 25, 1985, discloses that the FDA had drafted a proposed regulation that would have lowered the level to .1 ppm – a 100-fold reduction from the present level.

Public Citizen's Katherine Meyer, attorney for the plaintiffs explains, "the law clearly provides that food additives that cause cancer in animals must be banned from the food supply. Anything less than a total ban violates the law and deprives consumers of the protection Congress intended." Ms. Meyer also noted that "the FDA's unwillingness to abide by the law is particularly egregious in light of the fact that some manufacturers of decaffeinated coffee use safe alternatives, such as water, in the decaffeination process. Therefore, while prohibiting the use of methylene chloride will reduce the risk of cancer, it will not result in banning decaffeinated coffee from the market." "Unfortunately," she added, "consumers cannot tell which brands use methylene chloride simply by reading the labels." Commenting on the government's use of a "de minimis" exception to the anticancer provision of the law, Dr. Sidney M. Wolfe, director of Public Citizen's Health Research Group, asserts that "de minimis is disastrous – it would allow the Reagan Administration to continue to ignore the laws enacted to protect the public's health, in favor of protecting the profits of the food industry. Particularly since water decaffeination is available, this decision by the Reagan Administration does not square with its stated goal of a 50 percent reduction in cancer mortality by the year 2000."

FOUR VIEWS ON NUTRASWEET: 1. Most Scientists in Poll Doubt NutraSweet's Safety

BY MICHAEL SPECTER
THE WASHINGTON POST, JULY 16, 1987

More than half the scientists responding to a government survey on the use of aspartame, the enormously popular sugar substitute marketed as NutraSweet, voiced concerns over its safety, according to a Government Accounting Office report released yesterday.

Contained in dozens of products, including diet sodas, breakfast cereals and vitamins, the artificial sweetener is consumed by 100 million Americans each day. Even before it was approved for use by the Food and Drug Administration in 1981, the product was controversial. Critics charge that it might cause brain tumors, seizures, mood swings and headaches.

Because it has no medical expertise, the GAO made no recommendations in its report. It found that the FDA adhered to proper procedures by approving the sweetener. But the investigators also said "we cannot comment on whether important issues surrounding aspartame's safety remain unresolved."

Officials of the NutraSweet Company hailed the results, and so did the sweetener's strongest critics.

"The report provides additional evidence to add to the overwhelming consensus among scientists and health officials...that aspartame is safe and that the process through which it was approved was proper and adequate," the company said in a statement.

The company dismissed concerns voiced by scientists who were polled that the substance should have been put through more rigorous tests, in part because some of those questioned have not conducted research on aspartame.

The FDA has received more than 3,000 complaints about the powerfully concentrated artificial sweetener since it appeared on the market six years ago. Consumers complained of dizziness, headaches and nausea after eating or drinking products with NutraSweet in them.

"They never asked the right questions about what it does to brain function in humans," said Dr. Louis Elsas, director of medical genetics at Emory University Medical School. "They decided without data that you had to have enormous amounts of phenylalanine in your blood before it becomes a problem. We don't know that's the case."

Phenylalanine, an amino acid, is the principal ingredient in NutraSweet. Those who suffer from phenylketonuria (PKU), a genetic disorder, cannot process the amino acid properly. All products containing NutraSweet must warn against its use where PKU exists.

Elsas and other researchers say they believe that aspartame can do more damage over a long period of time than federal health officials have said they think.

FDA officials have said repeatedly that aspartame has been subjected to stricter tests than any product ever approved. They declined comment yesterday, however, because they had not yet received a copy of the report.

The GAO report noted that the FDA Center for Drugs and Biologics, which oversees testing of new drugs, originally recommended that NutraSweet should have the more stringent testing required for drugs, rather than foods. FDA officials overruled the recommendation.

The report also said that before it was approved in 1981 by former FDA Commissioner Arthur Hull Hayes Jr., three of five FDA scientists advising the commissioner told him that studies conducted by the manufacturer had not conclusively proved the sweetener was safe.

The FDA polled 69 medical researchers, some studying aspartame. Twelve said they had "major" concerns about aspartame safety; 26 said they were "somewhat concerned," and 29 said they were "very confident" of the product's safety.

2. LETTER TO THE EDITOR
Aspartame: What the Findings Show
BY THOMAS E. STENZEL

Executive Director,
International Food Information Council
Alexandria, Virginia
THE WASHINGTON POST, AUGUST 1, 1987.

An overzealous headline writer and a reporter who apparently read a different document joined to misrepresent the findings of a General Accounting Office report on the food additive approval process followed for aspartame ["Most Scientists in Poll Doubt NutraSweet's Safety," July 17].

The two-year investigative report by the GAO was requested by Sen. Howard Metzenbaum, a self-avowed critic of aspartame, to determine whether the FDA had acted properly in the regulatory process that led to aspartame's approval. The GAO also investigated how the FDA is currently addressing any safety questions.

The report seems blatantly clear: "Throughout aspartame's approval history, GAO found that FDA addressed safety issues raised internally and by outside scientists and concerned citizens.

"GAO believes that FDA's and other scientists' planned and ongoing research, and FDA's monitoring of adverse reactions, should provide FDA with a basis for determining what future actions, if any, are needed on aspartame."

In announcing the report, a press release issued by Sen. Metzenbaum's office selected one portion of the 100-page document to trumpet–a survey questionnaire sent by GAO to 96 researchers. Of 69 who responded, more than one-third had done no research on aspartame. Even so, the results are interesting.

Twelve persons indicated they had little confidence in aspartame's safety; 29 said they were very confident of its safety. But consider that another 26 agreed with the following statement: "I am somewhat concerned about the safety of aspartame; I am generally confident in the safety of aspartame."

Doesn't this mean that 55 out of 67 researchers, more than 80 percent, are at least "generally confident" in the safety of aspartame? Does this match *The Post*'s headline?

The GAO also specifically stated that its questionnaire results, while providing useful information on ongoing research, "may not be totally representative of scientific opinion."

Aspartame has been determined safe by the FDA, which has properly reviewed the scientific evidence at every stage according to the GAO (so let's put those allegations to rest). It has also been determined safe by 60 countries, the European Economic Community's Scientific Committee for Food and the Joint Expert Committee on Food Additives of the U.N. and the World Health Organization.

Finally, it has been reviewed and recognized as safe for the general population by leading independent medical authorities, including the American Medical Association's Committee on Scientific Affairs, the American Academy of Pediatrics' Committee on Nutrition, the American Diabetes Association and others.

Perhaps a survey of these leading health and medical authorities around the world would have led *The Post* to write a different story. But then again, perhaps not.

3. LETTER TO THE EDITOR
Doubts About Aspartame
BY JANET SMITH

Executive Director,
Aspartame Consumer Safety Network
Washington, D.C.
THE WASHINGTON POST, AUGUST 22, 1987.

It is interesting that such a large percentage of scientists surveyed by the General Accounting Office in its recent report on the safety of aspartame expressed doubts about the safety of this product [news story, July 17]. This survey followed public announcements by several highly regarded, independently financed scientists. A high incidence of seizures, birth defects and behavioral disorders in laboratory animals raises disturbing questions about how the product may be affecting humans.

It is not surprising that the industry, in reacting to this news in a letter to the editor from Thomas Stenzel of the International Food Information Council [Aug.1], has once again ignored the obvious and immediate public-safety questions raised by these scientific concerns. IFIC is a trade organization representing eight food companies that market aspartame-containing products. Mr. Stenzel conveniently omitted mention of the fact that the GAO report on aspartame "did not evaluate the scientific issues raised concerning the studies used for aspartame's approval or FDA's resolution of these issues, nor did we determine aspartame's safety." The only comment on safety in that report was the survey of scientists published at the report's conclusion. Mr. Stenzel would have preferred that *The Post* dwell not on the issue of safety, but on the issue of how well the Food and Drug Administration followed the approval process in the first place.

If, as the GAO found, the FDA followed the law in approving aspartame (if only the letter rather than the spirit of it), and the product does now produce serious adverse health effects in some of its consumers–as we believe to be the case–then there is a need to change the law.

4. *Think Before You Sweeten*

BY PAT CROCE
THE PHILADELPHIA INQUIRER, SUNDAY, MARCH 15, 1987.

Have you ever ordered the most fattening dessert on the menu and then requested an artificial sweetener for your coffee?

It's not unusual. In fact, using artificial sweeteners in combination with a high-calorie meal is an all-American habit. But the next time you grab a diet cola or pour an artificial sweetener over your cereal, you might ask yourself just how beneficial or safe these synthetic sweeteners are.

Unfortunately, there are no easy answers. Artificial sweeteners have been the subject of controversy since the early 1970s and have been scrutinized by scientists, federal regulators and consumer advocates. What is undisputed is that, even though they make up a billion-dollar-a-year business, non-caloric artificial sweeteners have not helped the national crusade for thinness.

In fact, studies show that Americans are fatter than ever, and there is no evidence to support the idea that artificially sweetened food helps us lose weight. Think about it: When was the last time you heard a thin person attribute the secret of weight maintenance to diet soda?

As concerned as we are with sweeteners, both natural and synthetic, most consumers are not aware of the differences between them. Here is a basic overview of some common sugars and substitutes.

Sucrose. Table sugar is often referred to as "sweet 16" because a teaspoon has 16 calories. People avoid it because it contains "empty" calories (that is, it doesn't provide vitamins, minerals or fiber), causes weight gain and promotes tooth decay.

Fructose. A simpler, and sweeter, sugar found in honey and many fruits.

Saccharin. This non-caloric substance, 300 to 500 times sweeter than sugar, was discovered in 1879 and is used mainly in table-top sweeteners and soft drinks. An estimated 70 million Americans – one-third of them under the age of 10 – use it regularly. Various studies on rats have indicated that large doses of saccharin could cause bladder cancer in test animals, and in 1977 Congress required that foods containing saccharin be sold with a warning label: "Use of this product may be hazardous to your health. This product contains saccharin, which has been determined to cause cancer in laboratory animals."

Aspartame. This is the generic name for NutraSweet. It is the most widely used of all sugar substitutes, probably because it has no bitter aftertaste like saccharin. A combination of methanol and the amino acids phenylalanine and aspartic acid, aspartame is 180 to 200 times sweeter than sugar. It is used in 70 kinds of products, including diet drinks, cereals, desserts and ice creams. The Food and Drug Administration approved it in 1981, but has received about 2,000 health complaints from users since.

Scientists have expressed concerns that aspartame may trigger migraine attacks, alter brain function and cause brain tumors. Other complaints have centered on dizziness, headaches, hives, visual impairment and seizures. The NutraSweet Co. of Skokie, Illinois, has said that research has shown its product to be safe. The FDA says that a person may safely ingest 50 milligrams of aspartame per kilogram of body weight daily (about eight cans of diet soda for a 120-pound adult), but does not require food manufacturers to list the amount of aspartame in their products.

It does, however, require that goods containing aspartame be so labeled, to protect babies with a condition known as PKU, which leaves them unable to metabolize phenylalanine and can lead to mental retardation. In addition, researchers at Leeds University in England found that aspartame increased feelings of hunger in some people, whereas sugar was found to decrease hunger and produce a feeling of fullness.

Sorbitol. This is a sweet, crystalline alcohol commonly used in sugarless gum. It has almost the same number of calories as sucrose (8 calories in sugarless gum with sorbitol versus 10 calories found in sugared gum), but it does not promote tooth decay. Sorbitol may cause some side-effects, including gas, bloating and diarrhea. Many fruits are a natural source of sorbitol.

Cyclamates. These were marketed under the trade name Sucaryl and used extensively in the 1950s and '60s. They were banned by the FDA in 1970, however, when questions concerning their cancer-causing potential were raised.

Unless you are a diabetic and have to regulate your blood sugar, you may want to review your use of sugar substitutes. Sugar-free doesn't necessarily mean that a product is better for you. It is up to you to decide what saving the 16 calories in a teaspoon of sugar is worth to you. Make moderation your goal when using artificial sweeteners. And if you always grab an artificially sweetened product by reflex, think, the next time, about what you are putting into your body.

Antibiotics

Squandering a Medical Miracle

SAM ZUCKERMAN

NUTRITION ACTION HEALTHLETTER, JAN./FEB. 1985, PP.8-11.

In the years around World War II, antibiotics were added to the kit bag of medicine. It was a stunning breakthrough, ushering in an era of cheap, effective treatment for a host of previously intractable bacterial diseases.

Today we are in danger of squandering this medical miracle. Some 40 years into the era of mass-produced antibiotics, health providers find all too frequently that the drugs in their arsenal don't work. Massive overuse of antibiotics – not only in health facilities, but also, apparently, on the farm – is breeding strains of drug-resistant microorganisms.[1]

As a result, many agricultural uses of antibiotics are under fire. Yet, when the antibiotic age was getting under way, these seemed like one more of the benefits brought by the new wonder drugs.

In 1949, Dr. Thomas Jukes, a research scientist with the American Cyanamid Corp., found that one of the new products had a startling effect on livestock. Baby chicks fed a mash containing traces of chlortetracycline, Cyanamid's new antibiotic sold under the name Aureomycin, gained 10 to 20 percent more weight than normal; piglets did even more spectacularly.

Jukes' serendipitous discovery changed the livestock industry in as far-reaching a way as antibiotics revolutionized medicine. Poultry, cattle, and hog farmers began adding antibiotics to feed, even when animals were not sick. Routine low-level doses of these drugs, called "subtherapeutic" because the amounts are below those needed to treat outbreaks of infection, became an integral part of a new, more intensive system of animal husbandry. Not only did the drugs help livestock grow bigger and faster, but they protected against infection, thereby allowing animals to be confined tightly together without falling prey to the diseases crowding fosters. Livestock production took an enormous leap in scale, and the factory-type farms prevalent today became common.

Antibiotic feed additives first went on sale in the early 1950s. By 1983, American farm animals were devouring some nine million pounds of the stuff. In that year, American Cyanamid, Pfizer, and other drug makers racked up some $271 million in sales of these products, according to the Animal Health Institute (AHI), an industry trade group. Today, close to half the antibiotics manufactured in the U.S. are fed to farm animals.

To this day, Jukes stubbornly defends his innovation. But the practice sends shivers down the spines of many other scientists and public health specialists. With increasing vehemence, they are warning that regular use of antibiotics in animal feed is a Sword of Damocles poised over penicillin, tetracycline, and other drugs used to treat human diseases.

"It's perhaps one of the most serious public health problems the country faces," declares Natural Resources Defense Council (NRDC) senior scientist Karim Ahmed. "We're talking about rendering many of the most important antibiotics ineffective."

It's a matter of basic genetic law. Whenever antibiotics are used, whether in animals or humans, they kill susceptible bacteria and spare those that carry resistance to the drugs. Without competition, the surviving microorganisms thrive and pass on the resistance to their offspring. And, as long as antibiotics are in the vicinity, resistant microbes have a powerful survival advantage over competitors and can spread rapidly.

No doubt, the selection of antibiotic-resistant bacteria frequently comes from use of the drugs to treat human diseases. Yet there is no "Great Wall" separating the livestock and human worlds. Critics of antibiotic feed additives point to solid evidence that resistant microorganisms can sometimes find their way from animals to people.

And that's not the whole story. It seems that the resistance traits themselves, riding on separate bits of genetic material called plasmids, can jump from one species of bacterium to another. Plasmids are independent of a cell's regular chromosomes. If two species of bacteria come in contact in a pig's digestive tract, for example, microbes living in the animal gut can transfer plasmids carrying a resistance factor to varieties of bacteria that cause human diseases.

For Illness Only

Tufts University microbiologist Stuart Levy, one of the world's top authorities on plasmid biology, argues that penicillin and tetracycline should be reserved for treating disease, and not used to promote growth in farm animals. "This major use of antibiotics is a tremendous contributor to the environmental pool of resistant bacteria. And it's a use that could be eliminated," he declares.

Unless all antibiotics are used more wisely and more sparingly, we're likely to face an "evolution of more and more resistant plasmids, these transferable resistances...that can move between and among bacterial species," warns Levy. "I envision it like an overflow. You're building up this pool of resistant plasmids, and they're slowly encroaching upon more and more areas."

Restrictions on penicillin and tetracycline alone might reduce the problem, but they wouldn't eliminate it. Microorganisms often carry resistance to a range of antibiotics. Consequently, special types of antibiotics used only on livestock may also select for resistance to drugs used to treat human diseases. But a clampdown on penicillin and tetracycline is regarded as a vital first step.

A year and a half ago, Levy, Ahmed, and more than 300 other scientists signed an NRDC statement urging the Food and Drug Administration (FDA) to ban subtherapeutic uses of the two drugs on livestock. Several European countries took this step about a decade ago, although in Great Britain loopholes in the law still allow wide use of antibiotics in animal feed.

FDA in 1977 also sought to outlaw subtherapeutic penicillin feed additives, as well as all such uses of tetracycline except those considered essential to preventing infection. Use of the two drugs to treat animal diseases, as well as continued subtherapeutic use of special livestock antibiotics, would have been allowed. But, under intense pressure from the "animal health" industry, Congress blocked the proposal and demanded more research instead.

For their part, the drug makers continue to defend the practice. Links between human antibiotic resistance and animal feed are "in the realm of theory," contends AHI spokesperson Steve Kimbel. "We have yet to see research data that indicate there is a true threat to human health from this practice. It's... been going on for 30 years," he argues. "If there were some threat to human health, there would be ample evidence of its existence."

AHI finds itself increasingly far out on a limb. No longer is the connection between animal antibiotics and human disease just a theory. Dramatic new research findings have altered the equation. Separately, each of these studies breaks vital scientific ground. Together, they make a compelling case of a threat to human health.

In 1982, Harvard medical researcher Thomas O'Brien and colleagues reported a new method for demonstrating that resistance to antibiotics circulates between animals and humans.[2]

O'Brien used a technique for "fingerprinting" the plasmids carrying antibiotic resistance traits in bacteria. Each type of plasmid displays unique biochemical traits, and O'Brien found a way to profile them. He then found that cattle in some 20 states and 26 people in two states were infected with antibiotic-resistant Salmonella bacteria showing nearly identical plasmid profiles. He concluded that the plasmids carrying antibiotic resistance must have passed between the animals and humans. Otherwise the profiles would not have matched so closely.

More recently, researchers from the U.S. Centers for Disease Control (CDC) in Atlanta found that most of the Salmonella food poisoning episodes between 1971 and 1983 that they investigated could be traced to food animals, rather than sources such as contaminated kitchen workers. [3] The CDC team was able to pinpoint the sources of 16 outbreaks of antibiotic-resistant Salmonella infection. Food animals were responsible for 11 of those outbreaks.

But it is another piece of medical detective work that has made the biggest waves. CDC epidemiologist Scott Holmberg and colleagues reported in The New England Journal of Medicine last September what may be the first direct evidence of human disease specifically traced to food animals given antibiotic feed additives. [4] In an accompanying editorial, Stuart Levy called the study the "important missing link" providing the first "clear association between subtherapeutic use and human disease."

Holmberg investigated the cases of 18 Midwesterners infected with a strain of Salmonella resistant to tetracycline and other antibiotics. He was able, directly or circumstantially, to trace most of the cases to meat from a single South Dakota cattle herd fed low doses of the antibiotic chlortetracycline. Three South Dakota victims were relatives of the farmer who owned the cattle, and the three had eaten beef from the herd. Beef from the same herd was shipped to several Minnesota supermarkets where eight other patients said they bought hamburger before becoming ill.

AHI slammed Holmberg's report for presenting no evidence that the beef cattle ever harbored drug-resistant bacteria. And it's true that samples of the meat were not available for laboratory analysis. But the researchers did obtain a specimen of Salmonella bacteria from a dead dairy calf that had lived on the farm next door to the cattle. Using O'Brien's profile technique, they discovered that the bacteria's plasmid fingerprint matched those of the human Salmonella cases.

A yet-to-be-published FDA study of food poisoning in the Seattle area suggests that poultry may also be a path by which resistant bacteria reach humans. Most commercial poultry get special animal antibiotic additives, but varieties of tetracycline are also frequently used.

Researchers found that almost half the cases they

examined of a particularly virulent intestinal disease caused by Campylobacter bacteria stemmed from eating chicken. What's more, about 22 percent of chicken and turkey samples from local markets showed Campylobacter contamination. About 30 percent of the bacteria from the birds and a similar percentage from the humans showed the same pattern of resistance to tetracycline.

Long-time critics of antibiotic feed additives say these studies have sewn up their case. NRDC now calculates that penicillin and tetracycline are implicated in more than 270,000 cases of Salmonella infection each year, including 100 to 300 deaths. "You're going to have an epidemic of untreatable stomach ailments, many of which will end in death, because the antibiotics being used are ineffective," predicts Karim Ahmed.

And on top of this is the mind-boggling risk that antibiotics will lose effectiveness against a broad range of other diseases – not just those transmitted directly by tainted food – as resistant plasmids jump from one bacterial species to another.

Public Policy Needed

The rapidly accumulating scientific evidence is bringing to a head the question of what the federal government should do. In November [1984], NRDC petitioned federal authorities to impose an immediate ban on the subtherapeutic use of penicillin and tetracycline. The group asked Health and Human Services Secretary Margaret Heckler, whose department has jurisdiction over FDA, to take the extraordinary step of invoking the "imminent hazard" provision of federal law to bypass the usual bureaucratic hurdles and put such restrictions into effect at once.

FDA rarely acts so boldly. But, if it were up to FDA, a prohibition would already be on the books. The agency continues to support its eight-year-old proposal barring subtherapeutic uses of penicillin and restricting similar uses of tetracycline. Lester Crawford, director of FDA's Center for Veterinary Medicine, is an outspoken advocate of a ban.

Congress, lobbied heavily by farm and drug interests, has been the main roadblock. In FDA's 1979 appropriation, the lawmakers directed the agency to hold off action on farm antibiotics until the National Academy of Sciences could review the scientific literature. But the academy's 1980 report, which contended that a hazard to human health was "neither proven nor disproven," failed to settle matters. Drug manufacturers, livestock interests, and their supporters on Capitol Hill seized on these inconclusive findings. Congress then instructed FDA to carry out more research along lines suggested by NAS and each year since has ordered the agency not to act until the studies have been completed.

Now the verdict is in. The Seattle food poisoning report completes the research mandated by Congress – and the case against antibiotic feed additives is stronger than ever. Yet, AHI is gearing up another lobbying drive and there are signs that some lawmakers may again try to throw a monkey wrench into the process.

In October, shortly after publication of the Holmberg study, 13 House Republicans and nine Democrats sent Heckler a warning "to scrupulously avoid any action that would lead to doubt or confusion over the safety of the nation's meat supply because of the use of [antibiotic] feed additives." And Jamie Whitten (D-Miss.), the powerful chairman of the House Appropriations Committee and thus far the key opponent of action on antibiotics, has given no indication of any change of heart.

Congressional critics of subtherapeutic antibiotics are also aroused, however. In the last session, Rep. James Weaver (D-Ore.) introduced a bill that would enact FDA's proposed rule into law. Weaver, a long-time supporter of organic agriculture and environment causes, intends to push the legislation again in 1985. Meanwhile, the Food and Drug Administration was scheduled to hold hearings in late January [1985].

This may be the year that the logjam on antibiotics finally gets broken. But, 1985 could just as easily go down as the year that – despite the direct evidence of a threat to health – agricultural and chemical interests once again succeed in blocking government action. It's no longer a question of who can produce the best science. What counts now is who can do the best sales job.

Those who wish to support a federal ban on subtherapeutic penicillin and tetracycline feed additives should ask their congressional representatives to vote for legislation enacting such a prohibition and oppose any effort to delay action. Members may want to write Rep. Jamie Whitten, Chairman, House Appropriations Committee, 2362 Rayburn H.O.B., Washington, DC 20515, asking that his committee allow FDA to proceed with regulations on antibiotics. Letters to FDA urging action would also be helpful. Write Frank Young, Commissioner, Food and Drug Administration, 5600 Fishers Lane, Rockville, MD 20857.

Footnotes:

1. The best popular explanation of antibiotics in animal feed is, Orville Schell, Modern Meat: Antibiotics, Hormones, and the Pharmaceutical Farm, Random House, New York, 1984.
2. N. Eng. J. Med., 307: 1, 1982.
3. Science, 225: 833, 1984.
4. N. Eng. J. Med., 311: 617, 1984.

Press Release:
Consumer Groups Call For Warning Labels On Poultry and Meat

CENTER FOR SCIENCE IN THE PUBLIC INTEREST
JULY 1, 1987

A coalition of 21 consumer and environmental organizations concerned about food poisoning today asked the Department of Agriculture (USDA) to require all fresh poultry and meat to be labeled with cooking and handling instructions.

In a letter to Food Safety and Inspection Service administrator Donald Houston, the coalition urged USDA to "act promptly" to require labels warning consumers about bacteria that can cause food poisoning. The coalition cited Houston's testimony last month before a congressional committee, where he supported labeling of poultry.

The groups suggested that the labels state:

"Notice: This food may be contaminated with harmful bacteria. Cook thoroughly. Wash hands, dishes, and implements with soap and water immediately after contact with raw product."

Michael Jacobson, executive director of Center for Science in the Public Interest (CSPI), said, "The first line of defense should be clean farms and processing facilities. But until we have that, we must activate the last line of defense: the consumer. Labels on poultry and meat would remind consumers when they are in the kitchen of the need for extra care."

On May 12 the National Research Council (NRC), an arm of the National Academy of Sciences, recommended labels on poultry. Houston endorsed the idea three days later at a hearing of the Senate Subcommittee on Oversight of Government Management.

The NRC study group said visual inspection of chicken carcasses fails to detect Salmonella, Campylobacter, and other bacteria that can produce fever, diarrhea, and vomiting. The illness may last two to seven days and is sometimes fatal.

According to USDA estimates, one out of three chickens is contaminated with Salmonella. Part of the reason for the high contamination rate is speedy new processing machines that spread germs by spilling feces onto carcasses. The NRC report linked contaminated poultry to a large percentage of the 4 million cases of Salmonella and Campylobacter infections reported each year.

All told, the number of food-borne illnesses comes to more than 20 million a year, with approximately 9,000 deaths.

Because several outbreaks of Salmonella infection have been traced to hamburger, the groups urged that the labels be mandatory for all meat, not just poultry.

Among the groups signing the letter to USDA were Center for Science in the Public Interest, Consumer Federation of America, American Agriculture Movement of Illinois, Government Accountability Project, Natural Resources Defense Council, National Coalition Against the Misuse of Pesticides, and Public Voice for Food and Health Policy.

Other groups were: American Federation of Government Employees, Center for the Biology of Natural Systems (Queens College), Coalition for Alternatives in Nutrition and Healthcare, Inc., Concerned Consumer League of Milwaukee, Farm Animal Reform Movement, Food Animal Concerns Trust, Grand Forks (N.D.) Food Co-op, and Institute for Alternative Agriculture.

Also, Institute for Local Self-Reliance, International Alliance for Sustainable Agriculture, National Consumers League, National Women's Health Network, New Alchemy Institute, and the New England Small Farm Institute.

CSPI, which led the coalition's effort, is also spearheading Americans for Safe Food, a grassroots effort to improve the safety of the food supply.

CSPI is a nonprofit organization supported largely by its 80,000 members nationwide.

AMERICANS FOR SAFE FOOD
Letter to
U.S. Rep. Charles W. Stenholm

April 17, 1987

Charles W. Stenholm
Chairman, Subcommittee on Livestock,
Dairy, and Poultry
Committee on Agriculture
U.S. House of Representatives
Room 1301, Longworth House Office Building
Washington, D.C. 20515

Dear Rep. Stenholm:

The following comments on federal beef and pork inspection practices are submitted in place of testimony at the subcommittee's April 8th hearing on the matter.

Americans for Safe Food is a national coalition of consumer groups, environmental groups, animal rights groups, farm groups, and others concerned about the state of this nation's food supply. We represent a growing demand for uncontaminated food in the marketplace. For that reason, we feel that the inspection of beef and pork could and should be greatly improved. Far more carcasses should be tested for residues of antibiotics, hormones, and other drugs that may have been used on the animal. Much more attention should be paid to the possibility of microbial contamination, through a testing program that would cover various stages of processing. A foolproof recordkeeping system should be developed to track any meat contamination to its source. All of this should be done with government money by government employees. The trend toward allowing industry to police itself should be halted and reversed.

On the question of residues of carcinogenic drugs or pesticides, we believe that there are no reliably safe levels. Though it may take years before it can be diagnosed, cancer begins with the alteration of a single cell, and several of the animal drugs now in use (nitrofurazone, furazolidone, and carbadox, for example) have caused cancer in laboratory animals.

Though more research is needed, residues from hormones administered to promote livestock growth could possibly be affecting the moods as well as the sexual development of American meat eaters, especially if the withdrawal periods are not observed or these powerful drugs are otherwise misused.

Antibiotic residues could possibly affect people allergic to such drugs in addition to reducing the effectiveness of antibiotics in the treatment of human illness. The National Research Council (NRC), in its 1985 report, *Meat and Poultry Inspection,* states that the number of meat samples containing illegal levels of antibiotic residues was "very high indeed" for the samples tested from 1979 to 1983 – a violation rate of 1.48 percent.

According to the NRC report, the U.S. Department of Agriculture (USDA) runs laboratory tests on only one percent of all carcasses. The authors concluded that such a small sampling is not enough to protect the consumer, and must be increased. Furthermore, carcasses are rarely if ever tested for residues of many commonly used animal drugs, even carcinogenic ones. This situation must be rectified. Existing residue limits and withdrawal periods, weak as they may be, should be enforced rigorously. Those who violate such regulations should be punished severely, to set an example for others who are tempted.

The growing problem of microbial contamination could be the sleeping giant of food-related political issues. Over the past 20 years, the number of reported cases of salmonellosis doubled, according to statistics from the Centers for Disease Control (from about nine cases per 100,000 people in 1965 to about 19 cases per 100,000 in 1985). Current estimates are of 2 to 4 million cases a year nationwide. And salmonellosis is just one type of food poisoning. Altogether, foodborne infections sicken an estimated 20 to 90 million people and kill some 9,000 people in this country every year. The problem is all the more serious because of the appearance of "supergerms" – drug-resistant organisms that grow in the intestines of livestock treated with regular subtherapeutic doses of antibiotics. Such doses are administered not only to prevent disease among overcrowded animals, but also to increase their rate of growth. The fatality rate for salmonellosis caused by "supergerms" is higher than for cases caused by nonresistant germs.

Old-fashioned visual inspections are not enough to detect microbial contamination. According to the NRC report, the present postmortem inspection system is not adequate to detect human pathogens that don't produce an observable lesion on the carcass. The growing threat of food poisoning clearly calls for a new and conscientious testing system for a scientifically based sample of carcasses from every slaughterhouse or processing plant.

With meat contamination problems on the rise, consumers are shocked to see Congress, under the leadership of this committee, and the USDA taking steps to reduce consumer protection instead of increasing it. Recent legislation to reduce the number of inspections in meat processing plants and the USDA's other discretionary inspection programs will reduce the opportunity for products with observable lesions and fecal contamination to be removed from the line. These concessions to industry should never have been granted without taking important steps to protect the consumer as well. If mandatory microbiological testing were required, for example, perhaps we could get along with fewer government inspectors.

Without meaningful testing of the wholesomeness of our meat supply, however, the limited protection offered by inspectors should be increased, not diminished. The "Total Quality Control" program and other discretionary inspection systems in which company employees are substituted for government inspectors further undermine the already eroded confidence consumers have in the meat inspection program. Company employees are under too much pressure to cut corners and pass unwholesome meat. Also, they could easily mislabel products or otherwise shortchange consumers. Although the frequency of these types of activity is unknown, allegations of such behavior were made by USDA inspectors represented by the Government Accountability Project. Some other recent changes, such as a rule allowing a feces-contaminated carcass to be washed off instead of trimmed, should be reversed.

Inspection and sampling procedures should be modernized as recommended by the NRC report and should include laboratory tests for pesticide residues in addition to tests for the residues of hormones, antibiotics, sulfa drugs, and other animal drugs. There should be no tolerance level for any substance known to cause cancer in laboratory animals. Much more research should be devoted to testing the metabolites of these chemical as to how they may affect human health over time and in combination with other such residues. Testing methods must be developed for the residues and metabolites of the 70 percent of approved animal drugs that cannot be detected at present. Laboratory equipment and procedures should be upgraded so that contaminated meat cannot be sold before the test results are available.

There should be a greater effort to prevent meat contamination in the first place, at every step of the birth-to-supermarket cycle. A modern system of tracking any contamination to its source, as recommended by the NRC report, could help prevent a recurrence by re-educating those responsible for the contamination and placing the economic burden of condemned carcasses on the owner or feedlot operator. This would include mandatory identification, traceback, and recall systems for all food animals.

In short, our position is better safe than sorry. The measures recommended above, many of which will require new training for inspectors and other USDA personnel, will be worth the extra cost, and in fact are necessary if the Department is to live up to its mandate to protect the consumer.

Respectfully submitted,

Michael F. Jacobson, Ph.D.
Executive Director
Center for Science in the Public Interest
1501 16th Street, N.W.
Washington, D.C. 20036

Hazardous Fish: The Raw Facts
The Need for a Mandatory Federal Fish Inspection Program

PUBLIC VOICE FOR FOOD AND HEALTH POLICY

Introduction:

Fish is the only flesh food which does not have a mandatory federal inspection program. While poultry and all meats are subject to formal mandatory inspection processes conducted by the U.S. Department of Agriculture (USDA), only a patchwork of federal and state programs partially addresses selected aspects of fish harvesting and processing.

The National Marine Fisheries Service (NMFS), the largest federal domestic fish inspection program, inspected only 13 percent of fish and shellfish consumed in 1986. And because this voluntary program is a division of the Department of Commerce, it is designed more as a marketing tool than a public health program. Further, of the edible fishery products Americans consume, roughly 64.4 percent are imported and receive limited federal oversight. The vast majority of fish reaching consumers' plates *lacks inspection by any governmental agency.*

The consequences of no comprehensive federal fish inspection program are severe – yearly, thousands of Americans become ill and some people die from eating contaminated or improperly handled fish and shellfish. Fish and shellfish can transmit microbiological contaminants causing acute illness, and long-term health problems can result from environmental contaminants. Seafood carries bacteria, viruses, toxins and parasites that are harmful to human health. Residues from pesticides and industrial wastes enter the food chain and accumulate in fish and shellfish subsequently consumed by humans. Improper or careless processing and packaging contribute to contamination problems.

Lack of a federal fish inspection program is especially serious because fish is becoming an increasingly important part of the American diet. Fish consumption has increased more than 11 percent since 1980. As Americans learn more about the links between high-fat, high-cholesterol diets and heart disease, they more often choose fish as a low-fat source of protein. In addition, reports indicating fish may reduce the risk of heart disease may have encouraged increased consumption. Americans ate a record 14.7 pounds of fish per capita in 1986.

Americans should not have to subject themselves to significant health risks by consuming fish and shellfish, one of the potentially healthiest sources of low-fat protein. Public Voice believes that the dangers presented by contaminated fish and shellfish warrant a comprehensive, mandatory federal inspection program. An adequate inspection program would help guard against contamination in a variety of ways. Rigorous oversight, a well-developed sampling program and federal certification of fishing vessels could minimize the amount of contaminated fish that reaches consumers. A recordkeeping system could ensure that particular species or harvesting areas associated with an illness outbreak would be quickly identified and products from the area or similar species would be removed from the market. A public education component could alert consumers to hazards as well as identify proper and safe methods of handling and preparing fish. Finally, an adequate program would include residue testing to prevent fish with dangerous levels of toxic substances from reaching consumers.

Current Health Threats From Fish:

Potential hazards to human health from contaminated fish are serious. According to the Centers for Disease Control (CDC) in 1982 (the latest year for which data is available), 24 percent of all food-borne illness outbreaks where the source could be traced were caused by contaminated fish. In 1985, almost half the seafood product recalls were due to canning defects, which can cause botulism.

A wide variety of microbiological hazards can contaminate fish and shellfish, often from plankton or water filtered from their living environment. (See Chart). For example, vibrio vulnificus is a serious problem in raw or undercooked seafood. During the late summer of 1987, three people died from oysters contaminated with vibrio vulnificus bacteria, according to officials at the Food and Drug Administration. The state of Florida receives reports of five or six deaths from this bacteria in seafood a year. In 1986, 20 people contracted cholera and one person died from eating cholera-contaminated shellfish, according to the Centers for Disease Control in Atlanta.

These documented cases of illness from food

contamination represent only a small fraction of the thousands of illnesses attributable to contaminated seafood yearly. The CDC estimates that illness outbreaks they report account for only one to five percent of all illnesses attributed to food.

Consumption of raw fish poses special problems. A study published in the March 1986 *New England Journal of Medicine* examined 103 food poisoning outbreaks in which 1017 persons became ill from eating raw shellfish. An accompanying editorial in the *Journal* noted the "alarming frequency" of outbreaks of shellfish-related illness, particularly associated with eating raw or partially cooked products. The editorial concluded that existing enforcement policies were inadequate to prevent contaminated mollusks from reaching the public.

In addition to these microbiological contaminants, pesticides and chemical residues pose a very real hazard to public health. These chemical residues resist degradation and are incorporated into phytoplankton, which is the first link in the food chain. The residues become concentrated in the tissues of fish and shellfish consumed by humans. Many of these residues, such as DDT, mercury and dioxin, have been linked to cancer and other health problems.

Current Federal Programs:

While it is clear that the problem of fish contamination is widespread, no current federal inspection program offers continual, comprehensive protection against contamination. There is no unified federal program of fish inspection and the vast majority of fish are not inspected.

The National Marine Fisheries Service (NMFS), a division of the Department of Commerce; the Food and Drug Administration (FDA); the Environmental Protection Agency (EPA) and state agencies all have limited programs monitoring fish contamination. These programs fall far short of providing sufficient oversight and assurance that the fish Americans consume is safe. Sample testing is grossly inadequate and many states provide only minimal oversight, making reliance on state programs unacceptable. The lack of a unified program, coupled with the failures of the existing programs, put the seafood consumer in a precarious position.

NMFS offers a voluntary inspection program primarily used as a marketing tool, and not a comprehensive monitoring program. Although the inspection involves checks for wholesomeness, safety and labeling, its primary purpose is to facilitate marketing and promotion. Seafood handlers can request periodic inspections from NMFS which may amount to merely a "lot" inspection which involves sample inspection of a product.

The NMFS program inspects less and less fish and shellfish each year, despite the fact that fish consumption is on the rise. In 1983 NMFS inspected 567 million pounds, which represents approximately 18.6 percent of the fish and shellfish consumed; in 1984 NMFS inspected 483 million pounds, approximately 15 percent of the fish consumed and in 1985 and 1987 this percentage was down to 13 percent or less than 443 million pounds.

FDA has oversight authority of fish and shellfish under the Food, Drug and Cosmetic Act, which requires that food traveling in interstate commerce be safe and wholesome. Because the number of and variety of foods covered under this act is so large, seafood receives only sporadic oversight. FDA officials estimate that a fish processing establishment receives inspection only once every one to two years. From April 1980 to April 1986, only eight regulatory letters were issued by FDA to processors of fish and shellfish products, notifying manufacturers that they

---

Consumer Tips For Preparing And Eating Fish And Shellfish Safely...

- Avoid eating raw or undercooked seafood. Eating raw shellfish carries the highest risk of acute illness. Many harmful bacteria are killed by adequately cooking the fish. Persons being treated for liver disease, cancer, diabetes, AIDS/ARC/HIV infection, or any disease which alters their immune system and people who have had stomach surgery are advised never to eat any raw fish or shellfish.

- To find out if local fish is contaminated call or write your state Department of Public Health. They often issue advisories on restricted areas for sports fishing and consumption.

- Do not consume raw oysters, clams, shrimp, lobster or any seafood from areas that are contaminated with either industrial chemicals or biological contaminants. Be particularly aware of waters near major industrial plants or those contaminated with "red tides."

- Keep all seafood chilled well, between 32 and 40 degrees Fahrenheit. All raw seafood contains some bacteria that can multiply quickly in temperatures above 40 degrees F.

- Do not store unfrozen fish for more than a few days. Fish and shellfish can contain harmful bacteria and will ultimately deteriorate.

- Carefully trim all guts and skin from a fish. Pesticides and other industrial wastes are stored in the fatty portions of a fish. Pour off any extra oil before eating the cooked fish as it may contain some chemicals. Choose leaner types of fish, such as flounder, perch or grouper.

- Purchase fish and shellfish from reputable dealers, as others may not have proper storage and handling facilities.

were in violation of the law and ordering immediate action.

In cooperation with EPA, FDA sets tolerances or action levels for poisonous or harmful substances which might be present in food products. These tolerances or action levels are set to determine limits on the amount of potentially harmful substances present in food products. However, FDA has identified tolerances or action levels for only *fourteen* compounds in fish, despite the fact that hundreds of substances found in waterways pose potential health threats. For example, no tolerance or action level has been set for dioxin in Great Lakes fish.

In addition to the lack of comprehensive standards for the numerous potential chemical hazards, *only a minute percentage of fish is tested for dangerous levels of residues.* For example, the Food and Drug Administration tested only 532 fish for chemical residues in 1985. And Americans consumed over 3 billion pounds of fish in that same year. Also, according to a Loma Linda University official, in 1984-85 57 million pounds of fish were landed in Los Angeles alone, but FDA sampled only 30 fish through mid-year 1985.

Furthermore, records are not adequate to trace samples back to their place of origin, making it difficult for public health professionals to conduct further testing or take action to limit contaminated products from reaching the public.

State Action Responding to Fish Contamination:

In response to the myriad of potential health hazards in fish, certain states have instituted local pro-

Organizations Supporting Mandatory Federal Fish Inspection

Public Voice for Food and Health Policy
Americans for Safe Food
Arizona Consumer Council
Center for Science in the Public Interest
Concern, Incorporated
Concerned Consumers League
Consumer Action
Consumer Federation of America
Detroit Consumer Affairs Department
Earth-keep
Empire State Consumer Association
Environmental Action
Friends of the Earth
Gray Panthers of San Francisco
Human Environment Center
National Consumers League
Natural Resources Defense Council
New York City Department of Consumer Affairs
Public Citizen Congress Watch

grams to reduce public health risks from contaminated fish. However, these programs are not comprehensive, there is little coordination between states and no federal oversight program coordinates these efforts.

Because EPA only has tolerance standards for a small number of toxic chemicals, there has been much controversy about the need for the EPA to take actions in conjunction with state efforts. For example, while the EPA has been petitioned several times and advised by local EPA officials to set tolerances or action levels for dioxin, which poses a particular threat in the Great Lakes region, no action has yet been taken. Consequently, many states issue separate advisories making specific recommendations of consumption levels. Two are Michigan and Wisconsin which conduct yearly surveys of state marine life.

In Wisconsin, potentially hazardous fish are divided into three groups: those that can be eaten in restricted quantities, those that should not be consumed by pregnant women or children, and those that should not be consumed by anyone. In 1987, pregnant women and children were advised not to eat approximately 27 varieties of fish caught from the waters of Wisconsin. In Michigan, individuals are advised to eat specified low quantities of fish from numerous rivers and lakes, and pregnant women, women planning to become pregnant and children are advised not to eat any of these fish.

Washington state publishes yearly information for commercial and recreational fishermen on "red tide" contamination, a microbial contamination found in plankton consumed by fish and shellfish. The shellfish in the Puget Sound area has been greatly affected by the spread of the red tide, which can currently be found throughout the sound. Red tides also have posed a problem for shellfish consumption in New York and Florida.

Some New York state waters tested had levels of chemical contamination in some fish that were higher than any state, federal or international human health guidelines. Because of these hazards, New York has closed some waterways to commercial fishing and has issued health advisories against sports fishing in these areas. For example, the state banned commercial fishing for striped bass in all New York waters in 1986. The problem is so severe that a New York State Department of Health advisory recommends eating no more than one meal per week of half a pound of fish taken from any water in the state.

The Solution - A Mandatory Fish Inspection Program:

Because of the scope of potential contamination in fish, and the current lack of a comprehensive monitoring program, Public Voice for Food and Health Policy strongly recommends that a revised, comprehensive and mandatory fish inspection program be established. There have been a number of proposals introduced in Congress over the last 20 years to establish a mandatory fish inspection program and hearings on the subject have been held in 1967, 1968, 1971 and 1974. Jurisdiction over inspection alternately has been proposed for the Department of Commerce, the Department of the Interior, FDA and the U.S. Department of Agriculture.

Any adopted program will have to cover a wide number of potential hazards, and several components are crucial to the success of a fish inspection program, including:

• certification of fishing vessels based on prescribed safety standards to ensure good construction, sanitary storage areas with adequate temperature control and protection of fish from sun and weather, and periodic, unannounced inspection.

• adequate microbial and chemical residues standards and a statistically designed sampling, reporting and recordkeeping system. This system should allow for effective and efficient tracing of sources of contamination, facilitating restriction of the amount of contaminated seafood that reaches the public.

• a record-keeping program that will allow contaminated fish and shellfish to be quickly identified and traced so that other sources of potentially contaminated seafood can be quickly identified and their introduction into the market curtailed.

• uniform state requirements and minimum standards assurance.

• strict, sanitary requirements for fish processing, handling, wholesaling and transportation of products and facilities.

• adequate enforcement authority for the agency with government oversight responsibility.

• inspection of imported seafood.

• a comprehensive public education program on the proper handling and preparation of fish and shellfish.

Public Voice believes that the USDA would be the agency that could best administer a fish inspection program. It has one of the largest research institutions, technical expertise to ensure against microbial contaminants and chemical residues, and significant expertise in the operation of comprehensive inspection systems because of its experiences with poultry and meat inspection. In addition, while FDA also has a mandate to protect the American food supply, that is the major mission of the USDA.

The American fish and shellfish industry also would benefit from a mandatory inspection system because contamination episodes have shaken public confidence and affected sales of uncontaminated products. For example, Maryland clammers have experienced declines in prices whenever Florida or New England clams are subject to red tides. An adequate inspection system could avoid this by building consumers' confidence that the areas of contamination would be precisely identified and isolated.

The technology is available to meet the public's needs for assurances of safe, uncontaminated fish and shellfish. Current programs subject consumers to unnecessary risks while fish consumption is on the rise. A mandatory Federal inspection system that gives one agency responsibility for its implementation would provide Americans with safer seafood, a potentially nutritious, low-fat source of protein.

Heavy Metal

NUTRITION ACTION HEALTHLETTER, MARCH, 1987, P. 13.

Question: What types of fish contain mercury? I'm concerned because my family eats a lot of canned tuna.

Gertrude P. Herich
Brooklyn, NY.

Doctor Tastebud: Large fish near the top of the food chain accumulate methyl-mercury, a toxic, organic form of mercury, in their fatty tissues. Swordfish are the worst offenders. Other predatory species, such as tuna, halibut, red snapper, perch, pike (especially walleye), bass, burbot and sheepshead, have also been found to contain the metal.

Exposure to mercury can cause a loss of peripheral and night vision, and muscle and speech coordination in adults. In infants and developing fetuses, mental retardation, poor coordination, and seizures may occur.

Government inspection of fish barely exists, and the vast majority of fish sold is never tested for chemical contamination. Because fetuses and infants may be ten times more sensitive to methyl-mercury than adults, pregnant and nursing women should limit their intake of the mercury-accumulating fish to two four-ounce servings per week. Since striped bass, catfish, and carp have been found to contain not only mercury, but also pollutants such as PCBs, DDT, and dioxin, pregnant and nursing women should avoid those species and swordfish altogether.

Microbiological And Parasite Hazards of Fish and Shellfish (Public Voice for Food and Health Policy)

Fish Source:	Toxins and Parasites:	Disease threat to Consumers:	Symptoms:
Most outbreaks caused by improperly preserved canned foods	C. botulinum	Botulism	Double vision, weakness, difficulty breathing, can cause paralysis by acting on the nervous system
Primarily bottom-dwelling shore fish caught near reefs, most commonly barracuda, red snapper, amberjack and grouper	Ciguatoxin	Ciguatera Fish Poisoning	Nausea, cramps, diarrhea, numbness of lips and tongue, reversal of hot and cold sensation, pain in extremities
Primarily tuna, mackerel, bonito and skipjack	Scombrotoxin	Scombroid Fish Poisoning	Flushing, headache, dizziness, burning of mouth and throat, bronchospasm, abdominal cramps, nausea, diarrhea
Pufferfish	Tetrodoxin	Fish Poisoning	Numbness, loss of perception, floating sensation
Mostly associated with eating raw or undercooked fish, such as sushi and sashimi	Anisakidnematode	Anisakiadis	Stomach pains, nausea and vomiting; may penetrate stomach wall
Mollusks[1], particularly those harvested from waters contaminated with sewage	Vibrio cholerae 01 bacteria	Cholera	Diarrhea; if case severe, can lead to dehydration and death
Mollusks, particularly those harvested in summer months (bacteria multiplies more rapidly in warm water)	Vibrio cholerae non-01 bacteria	Gastroenteritis	Diarrhea, vomiting, fever, abdominal cramps
Mollusks (particularly if eaten raw)	Vibrio parahaemolyticus bacteria	Gastroenteritis	Abdominal pain, diarrhea, nausea, headache, fever
Mollusks	Vibrio vulnificus bacteria	Septicemia	Abrupt onset of chills, fever and/or prostration (people with liver disease are at serious risk)
Mollusks	Salmonella bacteria	Salmonellosis	Nausea, fever, headache, abdominal cramps, diarrhea
Mollusks (particularly if eaten raw)	Hepatitis-A virus	Hepatitis	Appetite loss, vomiting, diarrhea, and abdominal pain
Mollusks	Norwalk virus	Gastroenteritis	Nausea, vomiting, diarrhea, and abdominal pain
Shellfish	Gonyaulax	Paralytic shellfish poisoning	Tingling numbness, burning sensation in the lips, tongue and face, muscular weakness, respiratory paralysis and gradual paralysis of extremities
Shellfish	Ptychodiscus	Neurotoxic shellfish poisoning	Tingling of extremities, reversal of hot and cold sensation, vomiting, diarrhea

Note[1]: Mollusks are a phylum of invertebrate animals including oysters, clams, mussels and snails.

Water, Water Everywhere
The Poisoning of America's Drinking Water
BY JONATHAN KING
PUBLIC CITIZEN, OCTOBER 1985

December 21, 1982 is a day that Richard Arlington of Fort Edward, New York remembers well. That night, as he and his wife were going to bed, his phone rang. Arlington went into the living room to answer it.

The caller was Brian Fear, a New York state health official. It wasn't exactly regular business hours, but then this wasn't an ordinary call. Fear warned Arlington to stop using his water immediately. "Shut your water off," Arlington remembers being told. "Do not use your bathroom. Do not take a shower. If you go to the bathroom, put the [toilet] lid down, flush it and get out." The official explained that there were 11,000 parts per billion (ppb) of TCE in his water.

"This didn't mean anything at all to me," the 70-year-old retiree continued. "I didn't know what TCE was. I didn't know what 11,000 [ppb] meant." But he knew that it was pretty serious if a state official had taken the trouble to call him that late at night.

Fear had a good reason to bother Arlington with his after-hours call. The level of TCE in Arlington's water was extremely high – more than 200 times the state's recommended exposure level of 50 ppb. And TCE, or trichloroethylene – a clear, colorless, widely-used solvent – is highly toxic. It can strip away skin oils, causing rashes and cracking of the skin. In high doses, it can cause headaches, nausea, dizziness and kidney problems. Furthermore, it has produced cancer in laboratory animals, and is suspected of causing cancer in humans as well.

The source of the water contamination is visible from Arlington's backyard: a General Electric (GE) plant that manufactures electrical capacitors. Over the years, chemicals from the factory soaked and resoaked the ground, seeping down into the groundwater. They then slowly snaked toward the wells on Stevens Lane, where Arlington's house is located. And Arlington's situation was hardly the worst. One of his neighbor's wells posted an initial reading of 50,000 parts per billion of TCE.

For the next several months Arlington and his wife had to drive 40 miles round trip to their daughter's house to take a shower. And Arlington, who suffers from heart problems, had to carry jugs of water from friends' houses.

Fortunately these were temporary inconveniences. Arlington's and his neighbors' plumbing systems were connected to the Fort Edward municipal water supply the following May. But the after-effects of the water contamination may last longer. The health of Fort Edward residents may be affected in years to come. Several residents, including children, now suffer from repeated bladder infections. One child has developed a stomach ulcer. They wonder if TCE caused any of these problems. It can probably never be proven.

The Next Great Domestic Crisis:

Most Americans turn on the tap to get a glass of water or step into a shower without a second thought. Unhealthy tap water is something we tend to associate with Third World nations. But from the backwoods of Maine to the corn fields of Nebraska to the lush islands of Hawaii, more and more people are finding that their water is laced with a variety of toxic contaminants, many of which are capable of causing cancer, birth defects, nerve and organ damage and other diseases at even low levels of exposure. In fact, we are finding that we can no longer take clean water for granted. As Lawrence Gladieux, a Florida highway patrolman who discovered his well was contaminated with the cancer-causing pesticide ethylene dibromide (EDB), said, "Water is very precious. But not until you are deprived of it do you realize that. We have got to guard it like it is gold."

But we have not guarded it. Instead, we have fouled it so badly in some areas that the pollution may be irreversible. When asked about the extent of the problem, one Environmental Protection Agency (EPA) official threw up her hands and said, "There are areas where so much crap went – what can you do?" And Democratic Rep. Mike Synar of Oklahoma, chairman of the House Government Operations Subcommittee on Environment, Energy and Natural Resources, warns that "the next great domestic crisis we may face as a nation is a water crisis and...its solution may be more expensive and more elusive than the energy crisis."

Although we have lived through several energy crises since the mid-1970s, the idea of a "water crisis" has not really sunk into the American consciousness. In fact, such talk sounds like hyperbole – until one begins to consider the magnitude of the threat to our water supplies.

The problem extends far beyond a few toxic waste nightmares like Love Canal. The danger is far more widespread than stacks of rusting 55-gallon chemical drums or huge waste ponds leaking dark, oily sludge. There are literally millions of potential sources of contamination that range from the neighborhood gas station to high-technology electronics plants. They include:

• More than 600,000 open and closed solid waste facilities, such as trash dumps and sanitary landfills, according to the Congressional Office of Technology Assessment (OTA). OTA says that up to 35,000 of these may be threatening ground water.

• 2.5 million underground tanks used to store gasoline and another 500,000 to 1 million tanks containing industrial chemicals. According to the EPA, up to 25 percent of these could leak, perhaps into underground water supplies.

• More than 180,000 pits and ponds – called surface impoundments – that contain liquid wastes from industrial operations, mining, oil and gas drilling, and farming. About 70 percent of the 26,000 ponds that contain toxic industrial chemicals have no barriers to prevent leakage.

• Pesticides and fertilizers applied to agricultural fields that have tainted wide stretches of the country. Public health officials have found pesticides in groundwater in more than 20 states.

Probably only a fraction of these sites will cause serious water contamination, but the damage already done is tremendous, particularly in the case of groundwater, which until a few years ago was considered relatively immune from chemicals.

Surface water – our streams, rivers and lakes – constitutes only about four percent of the water found in the continental United States. The remainder is located in porous layers of sand or gravel called aquifers, through which the water moves at a trickle, sometimes only a few inches a day.

Groundwater is not isolated from surface water; the two are intimately connected. Groundwater seeps into springs and contributes about 30 percent of the volume of the nation's streams, rivers and lakes. Rain and surface water, in turn, drain down through the earth and replenish groundwater.

Groundwater is an integral part of our lives and our economy. It provides drinking water for about half the population, including nine of every 10 rural residents. It also provides a substantial amount of the water for industry and agriculture. With 12 to 14 million wells sunk into the earth, our use of groundwater has more than doubled since 1950, and all indications are that we will become increasingly dependent on it.

Perhaps because groundwater is not visible, it was, until recently, considered a pristine resource to be tapped at will. By the late 1970s, however, the billions of tons of chemicals that had been dumped into the ground over the previous decades were beginning to show up with increasing frequency in drinking water supplies. A 1980 EPA survey found that 350 hazardous waste disposal sites had caused 168 cases of groundwater or drinking water contamination in 32 states, forcing the closing of nearly 500 wells.

Worse still, not only can chemicals easily soak through the intervening layer of soil and invade the water table, but once in groundwater the contaminants do not readily disperse, settle out or degrade. A leak of only one gallon of gasoline a day, for example, can render the groundwater supply for a town of 50,000 people unfit to drink. And in its cool, dark environment, groundwater may remain contaminated for hundreds of years. In 1980, then-EPA Assistant Administrator for Water and Waste Management Eckardt Beck warned, "Our groundwaters, long considered virtually pollution-free, are threatened by ruinous contamination."

Unfortunately, in the last few years the situation have become worse, not better.

The Extent Of The Problem:

Nobody knows how much of our groundwater is polluted by chemicals from agriculture, industry or everyday household use. The estimates range from less than one to nearly four percent. On the face of it, it does not appear that extensive, and government regulators, while admitting that it is a serious problem, sometimes tend to downplay the extent of the contamination. "The amount of contaminated groundwater is very small compared to the amount of groundwater that's available for use in the United States," EPA Administrator Lee Thomas said in a recent interview with U.S. News & World Report.

Not surprisingly, however, most of the contamination is in the most accessible, heavily used aquifers, and millions of people are affected. An EPA survey of 945 public water supplies completed in the early 1980s found that nearly 30 percent were contaminated with organic chemicals. Meanwhile, researchers at Cornell University found that nearly two-thirds of the nation's rural households were drinking water that failed to meet at least one of the EPA's drinking water standards. And given that the EPA has set standards for only a handful of the several hundred contaminants identified in drinking water thus far, that survey undoubtedly underestimated the extent of the contamination.

Certain regions of the country have been particularly hard-hit. New Jersey, where every major aquifer is affected by chemical contamination, has capped more than 1,000 wells since 1971 because of pollution. On Long Island, the Suffolk County Department of Health Services has warned more than 740 households since 1978 that their wells are laced with toxic chemicals. In Florida, which is particularly vulnerable to groundwater contamination because of its shallow aquifers and sandy soil, the pesticide EDB has been found in more than 700 wells. Dozens of municipal wells serving Miami have also been affected. And ongoing sampling in California shows that nearly 20 percent of the public wells tested had at least traces of organic chemicals.

Even sparsely populated states are not immune from serious groundwater contamination problems. In Idaho, radioactive and chemical contamination from a U.S. Department of Energy nuclear weapons facility has spread over some 50 square miles of the underground water table.

Unfortunately, we still do not know the full scope of the problem. There is still no comprehensive survey of the extent of groundwater pollution or systematic monitoring of drinking water for most chemical contaminants. The state agencies responsible for enforcing drinking water standards and safeguarding water supplies are often so understaffed they can barely respond to citizen complaints and the most obvious problems. According to the Congressional Research Service, "the majority of information relating to groundwater contamination remains anecdotal, scattered and poorly organized...."

The Failure Of EPA Regulation:

On the books, the agency has the tools to prevent contamination of our water supplies and to ensure safe drinking water, but in practice the EPA has been slow to set standards, hesitant about promulgating strong regulations and often reluctant to enforce existing regulations. For example, during the 10 years following the passage of the Safe Drinking Water Act of 1974, the EPA referred only 21 violations of the act to the Justice Department – even though tens of thousands of violations occurred each year.

But the failure to regulate hazardous waste effectively is perhaps the agency's greatest failing. In 1976, Congress passed the Resource Conservation and Recovery Act (RCRA), which gave the EPA the authority to ensure the safe handling and disposal of hazardous waste. The agency, however, was not equipped to deal with the problem.

"When we began to write regulations for the control of hazardous waste...,we did not know how much there was of it or how hazardous it might be, and we didn't know where it was going," former EPA Administrator William Ruckelshaus admitted in 1984. Another official, from the agency's Office of Solid Waste, recalled, "We didn't know what hazardous waste was, let alone where it should go. It was laughable."

By the early 1980s, the EPA had finally enacted regulations to ensure the proper disposal of hazardous waste. These were supposed to minimize the threat of hazardous waste leaking out of land disposal facilities into drinking water supplies. But the biggest problem with the agency's hazardous waste disposal regulations is that they do little to discourage the continued reliance on land disposal of hazardous wastes – even though the agency has admitted there is no way to prevent waste from leaking into soil and eventually reaching the underlying water table. A survey released in April by a House Energy and Commerce subcommittee found that 559 of 1246 facilities had "some indication of groundwater contamination."

Many land disposal facilities regulated under RCRA – even some opened after the passage of RCRA – are ending up on the EPA's list of hazardous waste sites requiring cleanup under Superfund. By the end of 1984, there were 45 RCRA facilities on the NPL. And until recently, 80 percent of our hazardous waste continued to be dumped into the ground, according to OTA.

Exasperated by the problems and delays that plagued EPA's handling of hazardous waste, Congress finally passed legislation last fall [in 1984] that greatly restricts land disposal. The law, the Hazardous and Solid Waste Disposal Act of 1984 (which amends RCRA) will gradually phase out the disposal of many types of hazardous waste over the next several years unless the EPA determines that such disposal is safe. All unlined or leaking waste ponds are to be closed or equipped with liners by November 1988.

The legislation was hailed "as the beginning of the end for unrestricted land disposal" by Richard Fortuna, executive director of the Hazardous Waste Treatment Council, which is composed of firms primarily engaged in competing forms of waste management, such as incineration.

As the restrictions on landfills and waste ponds become tighter, another form of land disposal – deep well injection – is becoming increasingly popular for the same reason that landfills and waste ponds are popular: it is relatively cheap and easy. But not everyone agrees it is safe.

More than 11.5 billion gallons of liquid wastes are now injected several thousand feet into the earth each year, according to the EPA's Office of Drinking Water, making deep well injection the current leading form of disposal for hazardous waste.

Industry maintains that underground injection is a safe form of disposal if properly managed. Proponents point out that the waste is shot thousands of feet below usable aquifers, and thus does not threaten drinking water supplies.

Although deep wells do not pose as widespread or as immediate a threat to drinking water supplies as problem-plagued landfills and surface impoundments, "when problems do occur," said Fortuna, "they are lulus." In December 1983, for example, the Ohio Environmental Protection Agency closed down several deep injection wells at a disposal facility owned by Waste Management, Inc., the nation's largest waste disposal firm, after a leak of 45 million gallons of hazardous waste was discovered. The leak went undetected for almost two years.

The Alternatives:

There is some good news. We do possess the technology and know-how to substantially reduce the threat of contamination that the nearly 300 million tons of hazardous waste we generate annually poses to our water supplies. Most wastes can be destroyed, recycled, reused or detoxified. A 1983 report by the National Academy of Sciences concluded, "There exists some technology or combination of technologies capable of dealing with every hazardous waste so as to eliminate concern for future hazards."

The bad news is that we simply have not used this technology. In 1981, less than one percent of hazardous waste was disposed of using methods that destroy or detoxify the waste, according to David Anderson, a soil scientist and hazardous waste management expert with K.W. Brown and Associates of College Station, Texas. And it has not been the result of a lack of alternatives. A 1984 study by the EPA found that there were 158 billion gallons of unused hazardous waste treatment capacity in the United States in 1981.

The major reason for our paralysis is that land disposal is still the cheapest and easiest method of getting rid of waste – even though the long-term costs are more expensive in terms of money spent on cleanup and in terms of the toll on human health and the environment. According to Anderson, it still costs only about one-fourth to one-half as much to dump waste in a land disposal facility as it does to incinerate it.

The Future Toll:

While we can try to quantify the cost of cleaning up the nation's hazardous waste and drinking water contamination mess, there is no way to measure the toll that it will take on the health of millions of Americans potentially exposed to toxic and cancer-causing chemicals in their water. That is because it is extremely difficult to prove that any particular toxic exposure led to a specific illness.

Many of the chemicals found in drinking water have been linked to cancer and to problems with conception and pregnancy. And according to current statistics, one out of four Americans will get cancer, perhaps one out of five American couples who wish to have children cannot conceive, and a large percentage of pregnancies end in spontaneous abortion or miscarriage. But what proportion of these problems is caused by water contamination remains a mystery.

EDB:

For Lawrence Gladieux in Florida for example, this means if a member of his family develops cancer he may never know whether exposure to EDB was in any way responsible. "My children could be killed as sure as they were shot down by a criminal and there is nobody to point a finger at," he said. "There is no way to prove whose finger is on the trigger."

Occasionally, however, a certain community or neighborhood suffers such a high rate of health problems that somebody takes notice. Such is the case of Woburn, Massachusetts, an old industrial town north of Boston, and of San Jose, a city in northern California.

Massachusetts:

In 1979, the Commonwealth of Massachusetts found a number of toxic, cancer-causing chemicals including TCE, benzene and chloroform in two of Woburn's wells. The wells were closed, but many residents think the damage has already been done. By 1979, a local citizens organization had confirmed at least 10 cases of Leukemia among Woburn's children over the previous 15 years. In 1984, researchers from Harvard University School of Public Health released a study concluding that the rate of childhood leukemia in Woburn from 1969 to 1983 was more than twice the national average. The researchers concluded that "exposure to [the two] wells...is associated with an increase in the incidence of childhood leukemia."

The study also found a "consistent pattern of positive associations" between the contaminated water and increased risks of infant deaths; ear and eye birth defects; and urinary, respiratory, neurological and skin problems among Woburn's children. (See "The Hypocrisy of J. Peter Grace" in Public Citizen's July/August 1985 issue.)

In January 1985, the California State Department of Health Services released a study that found that pregnant women living in neighborhoods south of San Jose, where the drinking water was contaminated with industrial chemicals, had twice as many miscarriages over a two-year period as women living in similar areas and drinking uncontaminated water. The health department also found a high rate of birth defects among children born to women drinking the contaminated water, but stopped short of linking the problems to the contamination.

Research for this article was provided by Essential Information.

Citizen Drinking Water Questionnaire
BUYER'S MARKET, MARCH 1987

If you want to know about the drinking water in your locale, ask your metropolitan or community water supplier to respond to these twenty questions.

If you get a written response, we would appreciate receiving a copy addressed to Drinking Water Project, P.O. Box 19367, Washington, DC, 20036.

1. What is the size and location of the population served by this water system?

2. What are the source(s) of the water supply? If surface water, name the source. If groundwater, give the number and capacity of wells.

3. What technologies are currently used in the treatment system? What changes, if any, have been implemented in your treatment system since 1974?

4. Do you conduct free tap water tests when a complaint or request is made? Do you offer consumer guides to home water testing services and home water filtration systems?

5. Give a record of compliance with federal monitoring and reporting requirements and drinking water standards for the past five years. List violations by total number and as a percentage of tests completed.

6. Where do you publish public notices of violations of maximum contaminant levels (i.e. contaminant violations)?

7. How long after violation of a contaminant's drinking water standard is a public notice published?

8. Have you ever tested for contaminants other than those covered by regulation?

9. If so, who did the testing?

10. What were the results of the tests? List contaminants detected and concentration levels.

11. Have wells or intake valves in the system ever been closed because of a contamination problem? If so, list the contaminant(s) that caused the closing, at what level the contaminant(s) was detected, and the date of the closing.

12. If a closing has occurred, what action, if any, was taken to remove the contamination?

13. Which treatment technology do you use, granular activated carbon or packed tower aeration? List date the system was installed, the amount of water treated (if all wells are not filtered) and annual operating costs.

14. If you do not use granular activated carbon or packed tower aeration, do you plan to install either or both?

15. Is chlorine used as a disinfectant? If so, at what point in the treatment process is it applied? What are the average Trihalomethane (THM) levels?

16. Are you aware that several non-THM chlorination by-products have been identified as potent mutagens in drinking water samples?

17. Have you completed the water corrosivity testing and distribution construction materials identification requirements set out by the Environmental Protection Agency in Federal Register 45, p.57346, August 27,1980?

18. If so, what are the corrosivity characteristics of your water supply and what construction materials are present in your distribution system?

19. The Safe Drinking Water Act of 1974 requires that finished water of surface water systems be tested once every year and once every three years for ground water systems. Since 1974, how many tap water tests have you conducted? Where do you take water samples? Are these tests made available to the public?

20. Do you have a pamphlet available detailing citizen's rights under the Safe Drinking Water Act that provides information on the supply and testing of drinking water, problems with contamination and which contaminants have been found?

Circle of Poison

from *Circle of Poison*, Chapter 1

BY DAVID WEIR AND MARK SHAPIRO
1981

Here pesticides are the dish of the day, and one swallows more poison than food. There is not even a living hen or pig, and lately even the children are often sick. Could it be that even the gift that God gives – children – we cannot have?

— Alfonso Castro, Colombian Farmer

There's no problem with the ban of DBCP within the United States. In fact, it was the best thing that could have happened to us. You can't sell it here anymore but you can still sell it anywhere else. Our big market has always been exports anyway.

— Executive, AMVAC Corporation

Small shops in Indonesia sell pesticides right alongside the potatoes and rice. The people just collect it in sugar sacks, milk cartons, Coke bottles – whatever is at hand.

— Lucas Brader,
UN Food and Agriculture Organization

Nearly half of the green coffee beans imported into the United States contain various levels – from traces to illegal residues – of pesticides that have been banned in the United States.

— Food and Drug Administration

This book documents a scandal of global proportions – the export of banned pesticides from the industrial countries to the third world. Massive advertising campaigns by multinational pesticide corporations – Dow, Shell, Chevron – have turned the third world into not only a booming growth market for pesticides, but also a dumping ground. Dozens of pesticides too dangerous for unrestricted use in the United States are shipped to underdeveloped countries. There, lack of regulation, illiteracy, and repressive working conditions can turn even a "safe" pesticide into a deadly weapon.

According to the World Health Organization, someone in the underdeveloped countries is poisoned by pesticides *every minute.* [1]

But we are victims too. Pesticide exports create a circle of poison, disabling workers in American chemical plants and later returning to us in the food we import. Drinking a morning coffee or enjoying a luncheon salad, the American consumer is eating pesticides banned or restricted in the United States, but legally shipped to the third world. The United States is among the world's top food importers and 10 percent of our imported food is officially rated as contaminated. [2] Although the Food and Drug Administration (FDA) is supposed to protect us from such hazards, during one 15-month period, the General Accounting Office (GAO) discovered that half of all the imported food identified by the FDA as pesticide-contaminated was marketed without any warning to consumers or penalty to importers. [3]

At least 25 percent of U.S. pesticide exports are products that are banned, heavily restricted, or have never been registered for use here. [4] Many have not been independently evaluated for their impacts on human health or the environment. Other pesticides are familiar poisons, widely known to cause cancer, birth defects and genetic mutations. Yet, the Federal Insecticide, Fungicide, and Rodenticide Act explicitly states that banned or unregistered pesticides are legal for export. [5]

In this book we concentrate on hazardous pesticides which are either banned, heavily restricted in their use, or under regulatory review in the United States. Some, such as DDT, are banned for any use in the United States; others, such as 2,4-D or toxaphene, are still widely used here but only for certain usages. …As we will discuss, even "safe" pesticides which are unrestricted in the United Sates may have much more damaging effects on people and the environment when used under more brutal conditions in the third world.

In the United States, a mere dozen multinational corporations dominated the $7-billion-a-year pesticide market. Many are conglomerates with major sales in oil, petro-chemicals, plastics, drugs and mining.

The list of companies selling hazardous pesticides to the third world reads like a Who's Who of the $350-billion-per-year [6] chemical industry: Dow, Shell, Stauffer, Chevron, Ciba-Geigy, Rohm & Haas, Hoechst, Bayer, Monsanto, ICI, Dupont, Hercules, Hooker, Velsicol, Allied, Union Carbide, and many others.

Tens of thousands of pounds of DBCP, heptachlor,

chlordane, BHC, lindane, 2,4,5-T and DDT are allowed to be exported each year from the United State, even though they are considered too dangerous for unrestricted domestic use. [7]

"You need to point out to the world," Dr. Harold Hubbard of the U.N.'s Pan American Health Organization told us, "that there is absolutely no control over the manufacture, the transportation, the storage, the record-keeping – the entire distribution of this stuff. These very toxic pesticides are being thrown all over the world and there's no control over any of it!"[8]

Not only do the chemical corporations manufacture hazardous pesticides, but their subsidiaries in the third world import and distribute them.

• Ortho: In Costa Rica, Ortho is the main importer of seven banned or heavily restricted U.S. pesticides – DDT, aldrin, dieldrin, heptachlor, chlordane, endrin, and BHC. Ortho is a division of Chevron Chemical Company, an arm of Standard Oil of California.[9]

• Shell, Velsicol, Bayer, American Cyanamid, Hercules and Monsanto: In Ecuador these corporations are the main importers of pesticides banned or restricted in the United States – aldrin, dieldrin, endrin, heptachlor, kepone, and mirex. [10]

• Bayer and Pfizer: In the Philippines these multinationals import methyl parathion and malathion respectively; [11] neither is banned but both are extremely hazardous.

The Ministry of Agriculture of Colombia registers 14 multinationals which import practically all the pesticides banned by the United States since 1970. [12] And in the Philippines, the giant food conglomerate Castle & Cooke (Dole brand) imports banned DBCP for banana and pineapple operations there. [13]

Pesticides: a pound per person

Worldwide pesticide sales are exploding. The amount of pesticides exported from the U.S. has almost doubled over the last 15 years.[14] The industry now produces four billion pounds of pesticides each year – more than one pound for every person on earth. [15] Almost all are produced in the industrial countries, but 20 percent are exported to the third world.[16]

And the percentage exported is likely to increase rapidly: The GAO predicts that during the decade ending in 1984, the use of pesticides in Africa, for example, will more than quintuple.[17] As the U.S. pesticide market is "approaching saturation...U.S. pesticide producers have been directing their attention toward the export potential...exports have almost doubled since 1965 and currently account for 30 percent of total domestic pesticide production," the trade publication *Chemical Economics Newsletter* noted.[18]

Corporate executives justify the pesticide explosion with what sounds like a reasonable explanation: the hungry world needs our pesticides in its fight against famine. But their worlds ring hollow: in third world fields most pesticides are applied to luxury, export crops, not to food staples the local people will eat. Instead of helping the poor to eat better, technology is overexposing them to chemicals that cause cancer, birth defects, sterility and nerve damage.

"Blind" schedules, not "as needed"

But the crisis is not just the export of banned pesticides. A key problem in both the industrial countries and the third world is the massive overuse of pesticides resulting from their indiscriminate application. Pesticides are routinely applied according to schedules preset by the corporate sellers, not measured in precise response to actual pest threats in a specific field. By conservative estimate, U.S. farmers could cut insecticide use by 35 to 50 percent with no effect on crop production, simply by treating only when necessary rather than by schedule. [19] In Central America, researchers calculate that pesticide use, especially parathion, is 40 percent higher than necessary to achieve optimal profits.[20]

In the United States the result of pesticide overuse is the unnecessary poisoning of farmworkers and farmers – about 14,000 a year according to the Environmental Protection Agency (EPA).[21] But if pesticides are not used safely here – where most people can read warning labels, where a huge government agency (the EPA) oversees pesticide regulation, and where farmworker unions are fighting to protect the health of their members – can we expect these poisons to be used safely in the third world?

An inappropriate technology

In third world countries one or two officials often carry responsibility equivalent to that of the entire U.S. EPA. Workers are seldom told how the pesticides could hurt them. Most cannot read. And even if they could, labels on banned pesticides often do not carry the warnings required in the United States. Frequently repacked or simply scooped out into old cans, deadly pesticides are often handled like harmless white powder by peasants who have little experience with manmade poisons.

But perhaps even more critical is this question: can pesticides – poisons, by definition – be used safely in societies where workers have no right to organize, no right to strike, no right to refuse to carry the pesticides into the fields? In the Philippines, for example, at least one plantation owner has reportedly sprayed pesticides on workers trying to organize a strike. [22] And, in Central America, says entomologist Lou Falcon, who has worked there for many years, "The people who work in the fields are treated like half-humans, slaves really. When an airplane flies over to spray, they can leave if they want to. But they won't be paid their seven cents a day or whatever. They often live in huts in the middle of the field, so their homes, their children, and their food all get contaminated."[23]

Yet the President's Hazardous Substances Export Policy Task Force predicts that the export of banned pesticides is likely to increase as manufacturers unload these products on countries hooked on the ag-chemical habit. "Continued new discoveries of carcinogenic and other damaging effects of many substances are probable over the next few years," predicts the task force. "In some cases, certain firms may be left with stocks of materials which can no longer be sold in the United States, and the incentive to recover some of their investment by selling the product abroad may be considerable."[24]

The genetic boomerang

The pesticide explosion also has a second built-in boomerang. Besides the widespread contamination of imported food, the overuse of hazardous pesticides has created a global race of insect pests that are resistant to pesticides. The number of pesticide-resistant insect species doubled in just 12 years – from 182 in 1965 to 364 in 1977, according to the U.N. Food and Agriculture Organization. So more and more pesticides – including new, more potent ones – are needed every year just to maintain present yields.

A circle of victims

But enormous damage is done even before the pesticides leave American shores. At Occidental's DBCP plant in Lathrop, California, workers discovered too late that they were handling a product which made them sterile. Elsewhere in the United States, worker exposure to the pesticides Kepone and Phosvel resulted in terrible physical and mental damage.

As part of our investigation into the "circle of poison," we looked at these example of how the manufacture of hazardous pesticides affects American workers. Since companies are allowed to produce pesticides for export without providing health or safety data, there is no way to be sure they are not poisoning their own workers in the process. In fact, there is abundant evidence that workers in the industrial countries are indeed suffering from their employer's booming export sales.

We talked with West Coast pesticide workers who complained of inadequate protection – and information – on the job. Even after two hot showers, one group explained, their hands still carried enough toxic residue of an unregistered pesticide that, when they stuck a finger in a fish bowl, the goldfish died.

These workers in pesticide manufacturing plants are the very first victims in the circle of poison. Add to them all the people who load and unload the chemicals into and out of trucks, trains, ships and airplanes; and those who have to clean up the toxic spills which inevitably occur. Then the total number of potential American victims of hazardous pesticide exports becomes very large. In addition, of course, there are the victims whose story is told in this book – third world peasants, workers and consumers – as well as everyone else in the world who eats food contaminated with pesticide residues. We complete the circle of victims.

To uncover the story in this book we have had to overcome powerful obstacles. The pesticide industry is a secretive one. The Environmental Protection Agency guards industry production data from the public, press and even other government agencies. The information made available often seems to defy meaningful interpretation.

We filed over 50 requests under the Freedom of Information Act in order to penetrate the industry's "trade secrets" sanctuary inside the EPA and other agencies. In assembling hundreds of tiny pieces of the puzzle, we studied trade publications and overseas magazines and newspapers for evidence of hazardous pesticide sales. In addition, we obtained import figures from a number of third world countries. Finally, we interviewed hundreds of people in industry, government, unions, environmental groups and international organizations. We corresponded with farmers, consumers and environmental groups in the third world.

The story told here is intended not merely to shock and to outrage. Its purpose is to mobilize concerned people everywhere to halt the needless suffering caused by pesticides' circle of poison.

NOTES
CHAPTER ONE

1. Proceedings of the U.S. Strategy Conference on Pesticide Management, U.S. State Rept., June 7-8, 1979, p. 33.
2. "Report on Export of Products Banned by U.S. Regulatory Agencies," H. Rept. No. 95-1686, Oct. 4, 1978, p. 28.3
3. Ibid.
4. "Better Regulation of Pesticide Exports and Pesticide Residues in Imported Foods is Essential," U.S. GAO, Rept. No. CED-79-43, June 22, 1979, pp. iii, 39.
5. Frances Moore Lappe and Joseph Collins, Food First: Beyond the Myth of Scarcity (New York: Ballantine, 1979), p. 145.
6. "New Pesticides Must Now Be Economic Winners," Chemical Age, Feb. 17, 1978; Dr. Jay Young, Chemical Manufacturers Association, telephone interview with authors, Oct. 1979.
7. President's Hazardous Substances Export Policy Working Group, Fourth Draft Report, Jan. 7, 1980, p. 6.
8. Dr. Hal Hubbard, telephone interview with authors, June 1, 1977.
9. Lappe and Collins, op. cit., p. 146.

10. "Listudo de Pesticides Registrados en el Departamento de Sanidad Vegetal," Ministry of Agriculture, Ecuador.

11. "Importacion de Pesticidas," Ministerio de Agricultura y Ganaderia, Costa Rica, 1978.

12. "Plaguicidas de Uso Agricola, Defoliantes y Reguladores Fisiologicos de las Plantas Registrados en Colombia," Ministerio de Agricultura, Colombia, June 30, 1979.

13. (see Lappe and Collins, op. cit., p. 60.)

14. Thomas O'Toole, "Over 40 Percent of World's Food is Lost to Pests," Washington Post, March 6, 1977.

15. Douglas Starr, "'Pesticide Poisoning Alarming,' says FAO," Christian Science Monitor, Feb. 1, 1978.

16. Ibid.; and Lappe and Collins, op. cit., p. 64.

17. "Better Regulation of Pesticide Exports and Pesticide Residues in Imported Food is Essential," U.S. GAO Rept. No. CED-79-43, June 22, 1979, p. 1.

18. Jeanie Ayres, "Pesticide Industry Overview," Chemical Economics Newsletter, Jan.-Feb. 1978, p. 1.

19. Lappe and Collins, op. cit., p. 41.

20. "An environmental and Economic Study of the Consequences of Pesticide Use in Central American Cotton Production," Final Report, Instituto Centro-Americano de Investigacion y Technologia Industrial (I.C.A.I.T.I.), Jan. 1977, pp. 149, 155, 161.

21. Lappe and Collins, op. cit., p. 67.

22. Osawa Yasuo, "Banana Plantation Workers Strike in the Philippines," New Asia News, May 1980, p. 7.

23. Dr. Lou Falcon, telephone interview with authors, May 21, 1979.

24. Agriculture: Toward 2000, U.N. Food and Agriculture Organization (FAO), Rome, July 1979, p. 82.

25. 5 U.S.C. 552, Freedom of Information Act.

Harvest of Unknowns:
Pesticide Contamination in Imported Foods,

BY SHELLEY A. HEARNE
NATURAL RESOURCES DEFENSE COUNCIL, 1984.

Executive Summary

Today, a significant fraction of the fresh vegetables and fruits consumed annually in the United States comes from foreign soils. In response to the concern that this produce may be dangerously contaminated with toxic pesticide residues, the Natural Resources Defense Council (NRDC) initiated an investigation into the threat posed by these imported foods. This NRDC report addresses the problems associated with commodities exported from developing countries and the federal government's inability to halt the flow of chemically adulterated food products over the U.S. border.

Imported food products pose a unique problem to the safety of the American food supply because it is uncertain whether safe agricultural pesticides and practices were employed during the growing season. Many food-exporting nations in the developing world lack strict pesticide regulations. In turn, third world farmers are often untrained in safe and proper agrichemical uses, leading to misuse by overspraying or by applying dangerous products that are banned or severely restricted elsewhere in the world. Hazardous pesticides are still manufactured by industrial nations such as the U.S., and shipped to these food-exporting nations. These banned substances are widely used in many developing countries because they are generally cheaper than safer alternatives and are readily available.

To help assess the extent of pesticide contamination on fruits and vegetables imported into the U.S. from developing countries, the Natural Resources Defense Council reviewed previously unpublished Food and Drug Administration (FDA) data on their import monitoring program. The FDA has not published the results from the last eight years of this food safety program. An analysis of the FDA's findings show that many of the major food commodities imported into this country arrive with over half of their shipments containing pesticide residues. Imported foods have twice as many food quality violations as do domestically grown items.

Many of the chemical residues detected on imported produce are cancer causing substances which have been banned from agricultural use in the United States. These include pesticides such as DDT, lindane, BHC, and endrin, all of which are suspected to cause cancer, birth defects and other adverse health effects.

To substantiate these findings, the NRDC conducted an independent testing program on coffee beans in 1983. Coffee beans were selected for this case study because coffee is one of America's most popular beverages and 98% of the coffee consumed in this country is imported. FDA's surveys have found that 30 to 50% of the imported green coffee beans have illegal residues. The results of the NRDC test, employing improved methods, found multiple illegal chemical residues on every sample taken. For example, one coffee bean sample from Brazil contained detectable levels of DDT, BHC, lindane, aldrin and chlordane, all of which are prohibited in the U.S. Indeed, these test results exemplify the pervasiveness of pesticide contamination in imported foods.

Present U.S. government programs and policies

are unable to prevent adulterated food products from being sold in the American marketplace.

Only a small fraction of the total imported food shipments are inspected for pesticide contamination. For example, in 1982, only 14 samples were taken out of the two billion pounds of oranges imported into this country. This is equivalent to one sample for every 200 million pounds of imported oranges.

When the government does check food stuffs for chemical residues, the standard multiresidue tests employed detect less than half of the pesticides that may have been applied to the fruits and vegetables. Due to the limitations of the government's testing program, imported foods can be cleared for sale without ever being tested for contamination from the actual pesticides used. For example, the FDA does not test imported food items for the U.S.-banned pesticide, DBCP, a carcinogen known to cause sterility in males. Yet DBCP is reportedly used extensively worldwide, particularly on bananas and pineapples.

Even when the FDA does detect a pesticide contamination violation, there is no guarantee that the fruits or vegetables will not be sold to American consumers. Laboratory delays are often quite lengthy so that fresh produce is allowed to leave the shipping docks before the lab results are available. In 1982, more than one-quarter of the violative shipments reached the kitchen table in this fashion.

The importation of pesticide contaminated food items is too extensive to be halted singlehandedly by the federal government's stop-gap measures. Certain key improvements within the food monitoring program are vital, but the problems associated with imported foods can only be eliminated by addressing the issue as a whole. The U.S. government, along with other countries, international organizations and corporations, must focus on preventing the misuse of pesticides in the developing world if a long range solution is to be found.

Avoiding Pesticides in Foods

For the many who are unable to grow their own food sources or must rely upon commercial produce in the colder seasons, the following food preparation and dietary techniques are recommended:

• wash all fruits and vegetables well with soap before eating;

• soak potentially contaminated food stuffs in a mixture composed of 1/4 cup of vinegar and one gallon of water for five minutes. Rinse thoroughly in cold water. The acidity of vinegar can catalyze the breakdown of certain pesticides;

• peel all root-type vegetables (e.g., carrots, turnips, and potatoes). Systemic pesticides can concentrate in the skins of these produce;

• avoid imported goods since many of these crops may have been treated with pesticides not regulated in the U.S.;

• by maintaining good physical fitness and a nutritious diet, pesticides are less apt to accumulate in the body. Dietary fibers can help eliminate pesticides in the digestive system, as is true with Lindane, a carcinogenic chemical.

Adopted from "How to Avoid Pesticides (Sometimes)" by Hugh Delehanty in Medical Self-Care, Fall, 1981.

NRDC's Coffee Bean Testing Program

During 1983, NRDC conducted an independent testing program on coffee beans using the services of an outside contract laboratory. Coffee was selected for this analysis for two reasons: (1) because of its widespread popularity as a beverage, and (2) because 98% of the coffee beans used in the U.S. are imported.

In the first phase of testing, twenty-three samples of green and roasted coffee beans obtained in New York City markets were analyzed for chemical residues by methods similar to the FDA's own laboratory procedures. Of the thirteen green beans sampled, over thirty percent were found to contain residues of chemicals banned for agricultural use in the United States. While the number of coffee beans sampled were limited, these results do concur with earlier FDA estimates. See Table 4(a) for results of NRDC's coffee bean pesticide residue tests.

In the second part of NRDC's testing program, four samples of green beans were analyzed by improved methods which increased pesticide detection capabilities by as much as ten-fold over the methods used by the FDA. Using these techniques, multiple pesticide residues were found on *every sample*. In one instance, Brazilian coffee beans (the type most used in U.S.) contained multiple residues of the carcinogens DDT, BHC, lindane, aldrin and chlordane. Indeed, these NRDC test results exemplify the pervasiveness of pesticide contamination in imported foods – a situation not as clearly illustrated through FDA's limited surveillance program.

NRDC's final phase of testing focused on roasted coffee beans. At present, the FDA assumes that "all pesticide residues found in… green coffee beans [are] significantly lower after roasting"[note deleted] and, therefore, pose no threat to the consumer. Yet, NRDC's test weakens the FDA argument.

Two samples of green coffee beans were analyzed during the roasting process at timed intervals in order to determine if residues would be destroyed by the heat. Contaminants were detectable at reduced levels afterwards, but in contrast to FDA's theory, the results also demonstrated that pesticide residues remained despite the roasting process. In fact, in the "before-and-after" test of one sample of Brazilian coffee beans, all of the original detectable level of DDD (0.002 ppm) – the toxic metabolite of DDT – remained on the bean after roasting. In the worst case scenario for its risk assessment calculation, the FDA claimed that 10% of the original level of DDT and its metabolite DDD would remain after roasting. It was this assumption which had allowed the Agency to conclude that "the current levels of pesticides in imported coffee beans do not pose a hazard to the consumer."[note deleted]

True – some of the coffee bean's pesticide residues and their breakdown products did in effect decrease below detectable levels after roasting. However, the uncertainty of the effects on pesticide residues during the roasting process, leads NRDC to question the FDA's disregard of established tolerance levels in coffee beans.

Imported Ethnic Foods
Exotic Fare But Buyer Beware
BY CHRIS W. LECOS
FDA CONSUMER, DEC. 1986-JAN. 1987

The scene is a familiar one for anyone who has ever shopped in an ethnic food store. The shelves are stacked with a variety of exotic, packaged goods one would not find readily – if at all – in a huge, neat, fluorescent-lit supermarket, and many of the products are labeled in the same foreign language being spoken by the store's operators and most of its customers.

There is an unknown number of such stores in the United States today. Some have been in operation for decades – a legacy from immigrants who brought to this country their distinctive cultures, including their food preferences. Others are more recent, sprouting in urban areas populated by the waves of new immigrants from Southeast Asia, the Middle East, Latin America, and other areas.

Many of these stores are small, family-run operations. Their patrons are not only immigrants and their U.S.-born offspring, but also other Americans whose appetites and curiosity are whetted by the cuisines of foreign countries.

Until recently, ethnic food stores generally did not attract much attention from local health inspectors. No outbreaks of food poisoning that could be linked to the products sold there had ever been reported, and the number of customers was usually small, compared to supermarkets and other high-volume food stores.

Now, this is all changing. In a recent memorandum to state health officials, FDA's Division of Federal-State Relations warned that random inspections by the agency in various parts of the country had uncovered an "unacceptably high rate of defects" in many imported, packaged foods that were earmarked for sale through ethnic stores and restaurants. To protect the public from the potential health hazards those defects pose, FDA has developed a 10-point program that includes fostering federal-state efforts and increased surveillance by the agency itself of foods coming into this country.

Although some of the defects uncovered by FDA were economic in nature and not health hazards, many pose potentially serious health risks. They include swollen, leaking and rusty containers; moldy and decomposed foods; dry foods in plastic bags contaminated by insects and vermin; other evidence of rodent and insect infestation in some establishments; the use of undeclared food additives; products short of the declared weight; and products labeled only in the language of the country from which they were shipped. In the United States, imported packaged foods – just like domestic products – must be labeled in English, and the stores that sell them must comply with all federal and state labeling and sanitation standards.

An FDA task force report issued in September 1986 summed up the problem this way:

"Foods in hermetically sealed containers (canned products) which have been insufficiently acidified, insufficiently heat treated, or which have the integrity of the container compromised pose a danger to consumers from pathogenic anaerobes, particularly Clostridium botulinum [the microbe that causes botulism]. Defective containers may also allow entry of other pathogenic bacteria or molds. Improperly labeled products usually are considered economic problems;

however, undeclared additives such as sulfites and Yellow No. 5 may cause adverse reactions in sensitive consumers [who are unaware of the additives' presence in the products]. Improperly labeled products also create serious problems in tracing their origin."

Surveys conducted by FDA field offices in recent years have uncovered defects in various food products from at least 30 countries. The largest number involved Oriental foods from various Southeast Asian nations. However, FDA food safety officials warned that, since similar violations could exist among products from other countries, consumers should check any imported products carefully. Manufacturing practices in many other countries, FDA officials point out, do not always meet the same standards as those followed in the United States.

The foods with the defects were packaged in crocks, glass jars, large metal cans, and plastic jars, pouches and bags.

FDA was first alerted to the problem after a random survey of 45 ethnic food stores in 1984 by Gordon K. Brown, regional food service consultant with the agency's Baltimore district office. Almost all of the stores featured products from Southeast Asian countries. The stores were located in the Washington, D.C., and Baltimore metropolitan areas, in Newark and Wilmington, Delaware, and in Philadelphia. Brown purchased a wide variety of defective products from the stores and prepared an exhibit that he presented to other FDA officials. His report graphically described his findings:

"Approximately 75 to 80 percent of these stores have products in cans and jars which are obviously unfit....These items can be found just by walking around the store, looking, and picking up a few items as a customer would. Some items are so bad that they can be seen from across the store, and can be smelled outside the store on the sidewalk....

"When making the purchases," he continued, "none of the clerks made comment on the condition of the food, except for one swollen can. The clerk said it was bad, and when I made no comment, he rang it up and put it in the bag. Another clerk exchanged two swollen cans for others not as bad, but made no comment...."

Brown also went into some stores whose owners either were unaware of how long some of the products had been sitting on a shelf, or had simply forgotten about them. "In one instance," he recalled during an interview, "I found products that had been on the shelf for seven to eight years."

Virtually the same defects reported by Brown were uncovered in a follow-up survey by FDA of 47 randomly selected stores in Wichita, Leavenworth, Junction City and Overland Park, Kansas; Des Moines and Ottumwa, Iowa; St. Louis, Missouri; and Omaha, Nebraska. Products from 12 countries were being sold in the stores.

Most of the unsafe food found in this survey was in cans and glass jars, and some containers had foods that were discolored, moldy and partially decomposed, including partly decomposed frozen fish. Other food safety violations included store operators who were illegally repacking, preparing and canning foods in their homes for resale; frozen and refrigerated foods that were being kept at inadequate temperatures; and other products that clearly should have been refrigerated, but were not. Most of the establishments also were infested with insects or rodents. It was also common to find vermin-infested rice and assorted dried food products.

FDA officials warned that consumers were also endangered by the common practice of repackaging some of these imported foods, particularly those originally in cans. Such foods are usually processed to inhibit growth of bacteria or heat-treated to destroy harmful microorganisms. But if a can is opened and the contents are improperly repackaged, there is the potential for the growth of harmful bacteria, including C. botulinum.

FDA's survey of the 47 midwestern establishments revealed that in nine stores, seafoods and meats were processed without the required equipment and utensils. In 15 others, various foods were being repacked illegally. In general, the four-state survey showed that the majority of the stores did not have the required equipment and utensils to process meats or repack foods, adequate refrigerator and freezer facilities, or equipment to keep the facilities clean. In over half the stores, FDA said some packages were labeled only in a foreign language, or there was no labeling at all to identify the product.

So far, FDA officials said, there have been no recorded instances of food-borne illness traceable to ethnic foods. "This may be due partly to the fact that some of the violative foods are highly acidic or contain other natural preservatives, such as sugar or salt, which discourage the growth of pathogens," according to Thomas L. Schwarz of FDA's Retail Food Protection Branch. However, many newly arrived immigrants also come from countries where unsanitary conditions are the rule rather than the exception, so "we may be dealing with an ethnic clientele that is not in the habit of reporting a food-borne illness," he added.

Some imported food wholesalers and retailers have been closed and products seized in various cities. However, FDA officials stress that enforcement alone will not get the job done. In many instances, regulatory officials are faced with serious language barriers and other problems that they do not encounter when dealing with other food companies.

Most domestic companies, FDA officials said, are familiar with federal and state health and sanitation requirements, and they are usually represented by large trade organizations, making it much easier for regulatory agencies to transmit information when problems arise. "It's easy in these cases to get the word out and to get products off the shelves, when necessary," Schwarz pointed out. But this is not the case with many ethnic food stores.

"Many of these ethnic stores are Mom and Pop operations which do not belong to any organization," he said. "There are no mailing lists for distributors, and no other sources of information for tracking the distribution of a product [in the event a food-poisoning outbreak occurs or a product recall is necessary]. The retail store operators often are people who don't speak English, who have no knowledge or familiarity with federal and state sanitation and labeling requirements, who operate by standards that may be acceptable in the countries from which they came, but which are unacceptable in the United States."

Some operators, FDA's surveys have shown, did not understand the potential hazards with perishable foods that clearly should be refrigerated. In other instances, according to the agency's task force report, operators "did not know why they could not process or sell red meats of unknown origin without proper inspection and/or equipment and utensils."

Many store operators could not identify the manufacturers or distributors from whom the products were received. Some operators receiving defective products also were apparently unaware of how to return and get credit for bad shipments. "When the operators were asked as to the disposition of questionable or damaged foods on the shelves, none could give an answer," the FDA report recounted. "Evidently, no guarantee or agreement was made with distributors or manufacturers for receiving credit for questionable or damaged food...."

In urging state and local regulatory agencies to intensify their surveillance efforts, FDA pointed out that the problems found in the foods were partly a result of "errors in processing, packaging, labeling and quality control in the country of origin" as well as from "mishandling in this country during storage, transportation, repacking and retailing." In some instances, the agency has questioned the adequacy of processing techniques and preservation methods employed in some countries – methods that are not believed adequate to inhibit the growth of microorganisms that could create health hazards for consumers.

FDA personnel have been to China and Taiwan to help improve manufacturing practices there, but such efforts are limited because of the large costs to the agency and because FDA can only inspect plants at the invitation of the country involved. Instead, FDA plans to concentrate on the examination of packaged foods at U.S. ports of entry. The agency also intends to work more closely with the U.S. Customs Service and the U.S. Department of Agriculture and to encourage state and local health agencies to intensify their enforcement actions.

FDA's 10-point program also calls for using educational materials in the native languages of store operators and customers. Also involved in the effort are national food organizations, including the Association of Food and Drug Officials, whose membership consists mostly of federal and state regulatory officials. FDA already has provided state and local regulatory agencies with specific sanitation and enforcement guidelines for inspecting ethnic stores.

However, despite the intensified efforts, some products will continue to escape detection at ports of entry and on the shelves of some stores. The "first line of defense," as is so often the case, says FDA, is the consumer, who should examine carefully any imported products for possible defects.

Pesticides: Better Sampling and Enforcement Needed on Imported Food

U.S. GENERAL ACCOUNTING OFFICE REPORT, SEPTEMBER 1986.

PRINCIPAL FINDINGS

Limited Sampling Performed

Given the large number of food shipments entering the United States each year that could contain illegal pesticide residues and the limited number of samples taken, FDA's pesticide monitoring program provides limited protection against public exposure to illegal residues in food. FDA annually samples less than 1 percent of approximately 1 million imported food shipments.

FDA's general sample selection criteria include (1) high-volume imports, (2) foods of high dietary significance, and (3) products with past pesticide residue problems. The extent to which these factors are applied depends on the individual knowledge and judgment of FDA inspectors at the various ports of entry.

Between fiscal years 1979 and 1985, FDA collected and analyzed 33,687 imported food samples and found that 2,056 (6.1 percent) contained illegal residues. A review of the samples taken in fiscal year 1984 indicates that a large percentage of these samples were high-volume imported foods, while many lower volume imported foods were not sampled. In addition, foods imported from many countries are not being sampled. For example, shipments from only 9 of 27 countries exporting cucumbers to the United States from 1983 through 1985 have been sampled. The country exporting the second largest volume of cucumbers to the United States as well as 16 other countries had not had their cucumber shipments sampled since at least 1978, according to available records.

Lack of Pesticide Use Knowledge Hinders FDA

FDA generally uses multiresidue tests that can detect many pesticides on a single sample rather than single residue tests that can only detect one pesticide on a sample. FDA has five multiresidue tests that individually can detect from 24 to 123 pesticides. In combination these tests can detect 203 pesticides, less than one-half of the pesticide chemicals available for use worldwide. FDA laboratories normally use only one multiresidue method for each sample.

To select the proper test, FDA should have information on pesticides actually used on food produced in foreign countries. Little such information is cur-

rently available. Better information could be obtained from (1) U.S. manufacturers who export pesticides to countries that export food to the United States, (2) importers of food, if required to certify which pesticides were applied during food production, (3) a commercially available data source, and (4) cooperative agreements with foreign countries that export food to the United Sates. FDA is now in the process of obtaining commercially available data but will not know the impact of this data until later.

Deterrents Against Adulterated Shipments Not Used

FDA's policy requires importers to maintain all sampled shipments intact until the agency determines that the product is free of illegal pesticide residues. In practice, however, FDA permits importers to release the majority of sampled shipments to U.S. markets to allow consumers to receive fresh fruits and vegetables before they spoil...

Of 164 adulterated samples that GAO reviewed, 73 were not recovered and are presumed to have been consumed by the public. FDA recommended against damages in 52 of the 73 cases.

GAO was able to document only eight cases where importers were assessed [monetary] damages. Damages in six cases had not been collected a year after being assessed. Thus about 45 percent of the adulterated shipments are reaching consumers with few importers paying damages...

In order for the public to be protected from adulterated shipments and for the monitoring program to be an effective deterrent against such shipments, GAO believes that all importers of shipments determined to be adulterated should be assessed damages when the adultered food is not removed from the marketplace.

Recommendations

GAO recommends that the Secretary, Department of Health and Human Services, direct the FDA Commissioner to:

- redirect sampling coverage to a wider range of imported foods and countries and
- consider several options for obtaining additional information on pesticides actually used in foreign food production and to test for these pesticides. . .

Is Irradiated Food Safe or Necessary?

HRG HEALTH LETTER, MARCH/APR. 1985 AND NOV./DEC. 1986.

You may have heard or read about recent actions by the Food and Drug Administration to permit irradiation of pork products to control trichinosis and of fresh fruits and vegetables to control insects and inhibit growth. Unfortunately, many of the news accounts of these two FDA actions have been devoid of any significant discussion of the potential hazards of food irradiation. In addition, most of these accounts have failed to critically examine the claimed benefits of this technology. The Public Citizen Health Research Group is trying to present this side of this story.

Although food irradiation was previously approved by the FDA for use on certain commodities (wheat, white potatoes, dried spices and vegetable seasonings, and dry or dehydrated enzyme preparations), it was not until the proposal to allow irradiation for insect disinfestation and inhibition of growth and maturation of fresh fruits and vegetables that irradiation of a sizable portion of the American diet was at hand. This proposal, which was made public on February 14, 1984, would allow the irradiation of fresh fruits and vegetables at doses up to 100 krads. Doses up to this level could be used to delay ripening of fresh fruits and vegetables and to disinfest food of insects.

In the announcement of this proposal, FDA stated that it also planned to "delete the existing labeling requirement for retail packages of irradiated foods in the current regulations." This was based on FDA's belief that "any changes in food [from irradiation] are of no safety concern at the proposed doses and because the agency is not persuaded that special labeling is necessary." The Agency did, however, ask the public to submit comments on this issue.

FDA received thousands of comments from the public, the majority from consumers opposing the FDA proposal. However, on December 12, 1985, the Department of Health and Human Services, FDA's parent, announced that it intended to issue a final regulation allowing irradiation of fresh fruits and vegetables, subject to the review of the Office of Management and Budget (OMB). The FDA also announced that it was going to require a retail label that stated that the irradiated food was "picowaved."

Until FDA publishes a final regulation in the Federal Register, which will not be done until OMB has completed its review, the irradiation of fresh fruits and vegetables is not permitted. The irradiation of pork products, however, is another story.

On July 22, 1985, FDA announced that it had decided to allow the irradiation of fresh or previously frozen pork products to control the Trichinella spiralis parasite, the bug that causes trichinosis. This action came in response to an industry petition that had been filed one year earlier. The decision on retail labeling, however, was in the hands of the U.S. Department of Agriculture (USDA). Although USDA has said that labeling decisions would be made on a case-by-case basis, it has indicated that at the present time they will require that the labels include the word "irradiated" rather than the FDA's misleading term "picowaved."

Consuming Irradiated Food Has Not Been Shown To Be Safe

Despite the green light given to food irradiation by the FDA, the Public Citizen Health Research Group has serious reservations about the safety of consuming irradiated foods. This concern is not allayed by the FDA approval since that approval was not based on animal studies designed to assess risk but rather was based on a calculation of the amount of chemical changes induced in the food by the irradiation process.

When the FDA announced its proposal to allow irradiation of fruits and vegetables, it described the review of the safety of irradiation conducted by an internal agency task force, the Bureau of Foods' Irradiated Foods Committee (BFIC). This committee determined that the safety of irradiated food exposed to 100 krads of radiation or less could be presumed based on the "low level of total URPs" that would be created at such doses. URPs, or unique radiolytic products, are "new chemical constituents in the food...generated by the irradiation process." The same rationale was used in approving irradiation of pork. Thus, the BFIC, and FDA subsequently, *did not rely on any tests of toxicity or carcinogenesis conducted in animals to conclude that irradiated food was safe for human consumption.*

Perhaps the main reason FDA did not rely on long-term animal tests to assess the safety of irradiated food is that very few such tests exist. In addition, the animal tests that have been done were largely rejected by FDA as providing an inadequate basis for making a determination. An internal FDA review of 413 studies on the toxicity of irradiated foods found that 344 (84 percent) were either inconclusive or inadequate to demonstrate either safety or toxicity. Of the remaining studies, 32 indicated adverse effects and 37 appeared to support safety. The FDA memo states that "[o]n detailed examination of these 69 studies [only] five studies (one percent of all studies reviewed) appeared to support safety."

One recent set of studies that was not included in this review was carried out under the auspices of the USDA. This study, which was actually 12 different studies, examined the effect of feeding irradiated chicken to several animal species. One of these 12 studies found that fruit flies fed irradiated chicken had a statistically significant dose-related increase in the rate of death of their offspring compared with flies who were not fed irradiated chicken.

In another of these studies, a long-term study involving mice, the researchers found that mice fed irradiated chicken had a greater incidence of kidney disease than mice fed unirradiated chicken. In reviewing these findings, the group that did the work, Ralston Purina, concluded that "While no single finding from the study is highly illuminating, a collective assessment of study results argues against a definitive [sic]

conclusion that the gamma-irradiated test material was free of toxic properties."

The results on kidney damage are consistent with another study in which kidney damage was found in rats fed irradiated food. In that study the authors stated that "[t]he severity of these changes was directly dependent on the dose of irradiation of the food products."

Another study found testicular damage in rats fed irradiated food. Studies in both malnourished children and monkeys have demonstrated that consumption of freshly irradiated wheat may lead to an increase in abnormal white blood cells, a condition known as polyploidy.

Other studies have found that animals fed irradiated food are more likely to experience chromosomal damage. After reviewing these and other studies on this issue, Dr. Jonathan Ward, an expert in genetic toxicology at the University of Texas, made the following statement in a letter to FDA Commissioner Young:

"these studies as a group fail to either prove or disprove the existence of a hazard from the use of irradiated food. It appears possible that unstable mut-

---

What Has Been Done to Fight Food Irradiation?
HRG HEALTH LETTER, NOV.-DEC. 1986

The FDA's approval of food irradiation has prompted outcries from legislators and other concerned citizens who consider the FDA's action to be overly hasty and scientifically unfounded. In May 1986, the Health & Energy Institute and the Environmental Policy Institute - two public interest groups concerned about the health effects of radiation - petitioned the FDA to revoke its food irradiation regulation and hold a public hearing. And Rep. Douglas Bosco (D-CA) introduced the Food Irradiation Safety and Labeling Requirement Act to the House of Representatives (HR 4762). This bill, if passed, would prohibit the FDA and USDA food irradiation regulations from going into effect.

Bills have been introduced into the 100th Congress in both the House and the Senate concerning food irradiation. The Senate Bill, introduced by Senator George Mitchell of Maine, is called S.461. The House Bill, again introduced by Representative Doug Bosco of California, is called HR956. Both bills call for more research before food irradiation is allowed and extending the labeling requirement too irradiated ingredients used in restaurants and in processed foods. Write your senators and representatives urging them

to co-sponsor these bills. Banning food irradiation would be an even better idea.

In addition, a number of states - including Washington, Minnesota, New York, Vermont, and California - have either introduced or passed anti-food irradiation legislation with stricter labeling requirements than the FDA's regulation.

What Can You Do?
• Write to your Congresspersons to support HR 4762 (the Food Irradiation Safety and Labeling Requirement Act).

• Write to your Congresspersons to oppose HR 696 and S. 288, two federal food irradiation bills that would use your tax dollars to fund food irradiation technology and promote irradiated food to the public, and prevent states from passing stricter labeling laws.

• Express your opinion to government agencies such as the FDA, NRC, and USDA.

• Avoid eating or buying irradiated foods whenever possible. This will send a signal to food producers that consumers are not willing to be industry guinea pigs.

• Organize a local citizens' group to press for improved state labeling of irradiated foods.

agenic products may be produced in foods by irradiation....It is difficult to believe that adequate tests of the effects of radiation on food can be conducted without individually evaluating the components of the foods for which radiation sterilization is intended."

The Benefits of Irradiation Are Greatly Exaggerated

Irradiation has been touted by its proponents as an answer to a range of problems, including reducing reliance on pesticides and extending the shelf-life of fresh fruits and vegetables. However, careful examination of these claims reveals that the supposed benefits of irradiation are more imaginary than real.

In looking at how far irradiation can go in reducing our dependence on pesticides we must first consider that irradiation is obviously not a substitute for the seemingly infinite variety of pesticides that are applied to our food crops while they are still growing. Thus, irradiation can only substitute for pesticides that are applied after the commodity is harvested.

Part of the impetus for the renewed interest in food irradiation was the ban on the use of the fumigant ethylene dibromide (EDB). This chemical was used to rid various commodities, including citrus fruit, papaya and grain, of insects. Irradiation is now being touted as a substitute for these uses. However, there are a variety of other non-pesticide techniques for coping with the problems previously handled by EDB.

On citrus fruit EDB was used to kill various fruit fly species. However, this use called quarantine treatment, was only necessary on citrus shipped from one citrus growing region to another (e.g. Florida to California). It was not used on Florida citrus shipped to the Northeast or Midwest because fruit flies do not pose a danger to commodities in those regions.

Irradiation is not, however, a practical substitute for EDB. As researchers for the Department of Agriculture stated in a report issued in 1981:

"While research has established that irradiation is a potential treatment for commodities infested with insects, research on the tolerance of fresh fruit to irradiation has not been as promising, especially as regards citrus and mangoes. Our studies have indicated that adverse effects result at the relatively low doses required for quarantine treatment..."

USDA went on to say:

"Research has demonstrated that the dose of gamma irradiation required for fruit fly quarantine treatment could be less than 25 krads. But the necessity of treating large quantities of fruit and the desirability of treating already packaged fruit may mean that, to give a minimum dose to all fruit, the maximum dose might have to double or triple the minimum."

At such large doses, grapefruit would suffer significant damage, including skin pitting, scalding, and decay. In addition, irradiation may affect the taste of pasteurized and fresh juice and fruit sections. Other citrus fruits, such as tangerines and oranges, may be even more sensitive to peel damage from irradiation.

Various alternatives to EDB, other than irradiation, have been tried and appear promising. Grapefruits shipped to Japan can be disinfected by first exposing the fruit to 60 degrees Fahrenheit for seven days and then cold storage at 34 degrees Fahrenheit for 19 days. Since the 26 days necessary for the full treatment coincides nicely with the 22 days required to transport the fruit to Japan by boat the treatment can take place on board. The process appears to work well and has been accepted by the Japanese government as a substitute for EDB. USDA and the state of Florida are cooperating in the development of such a program for that state.

USDA is also investigating a method of killing fruit flies used before the use of EDB. This technique, a vapor heat treatment, involves a combination of high temperature and moisture. The technique does not damage the fruit and is applicable to a wide variety of commodities. It is already approved for use on several Hawaiian commodities, including papayas, bell peppers, pineapples, and tomatoes, and because it was used prior to EDB, could be used on produce from other regions of the country with little or no additional testing required by USDA.

Other techniques, such as fruit fly control programs, have proven to be effective alternatives to EDB. Since 1981, an effort has been underway to track the Mexican fruit fly population in the Rio Grande Valley of Texas. The program attempts to detect increases in the fly population and to manage the population by releasing sterile male insects. This approach allowed citrus to be shipped without EDB fumigation between October 1982 and March 1983, as well as this winter. Research is also progressing on techniques to detect fruit fly larvae in fruit so that infested fruit can be removed from shipments.

Apple growers in the state of Washington have been hoping that irradiation will help them penetrate the Japanese market. They believe that irradiation will eliminate the coddling moth from their produce, a pest that poses a threat to Japan commodities. However, even if irradiation is feasible for this purpose, it will not help these growers since the Japanese government will not accept irradiated produce.

The use of irradiation on store grain has also been proposed, although only about two percent of the stored grain in the U.S. was fumigated with EDB. For much of the grain no fumigants were necessary since the grain is exposed to cold temperatures which adequately prevent infestation. However, for the grain that does require treatment to destroy insects there are

Food Irradiation:
AMA vs BMA
HRG HEALTH LETTER, JUNE 1987.

Just as people are becoming more aware of the need to eat healthier food, more fruits and vegetables, along comes a serious threat to this important part of the food supply: FOOD IRRADIATION. At best, this scheme of zapping food with enormous doses of radiation is inadequately tested and some tests that have been done show deterioration of nutritional value. At worst, there is a well-founded fear that even though the irradiated food is not radioactive, it may cause genetic damage to those eating it.

In the United States, the Food and Drug Administration (FDA) has approved irradiation of fruits and vegetables with levels as high as 100,000 rads (equivalent to 4 million chest X-rays).

Sensing that consumers might not buy such foods, the food-zapping industry pushed for obscure labeling ("PICOWAVED"), some segments even wanting no labeling.

In the United Kingdom, however, food irradiation has been rejected because of a lack of evidence about its safety.

Of interest is the additional contrast between the views of organized medicine in these two countries. The British Medical Association (BMA), aware of the fact that although irradiation can kill bacteria infesting food, recognizes that irradiation does not inactivate the toxins already produced by the bacteria before they were killed. Dr. John Dawson, director of the BMA's professional and scientific division, said, "Radiation can also cause changes in food affecting its quality." Dr. Dawson went on to say, referring to genetic damage in animals induced by eating irradiated food, that "Further studies are needed before the potential benefits of radiation for the consumer, as opposed to the food manufacturers, can be assessed. We have to balance the benefits against the risks. We don't know enough yet to be certain that it is safe." Finally, the BMA said that the British Government's Ministry of Agriculture, Fisheries and Food Advisory Committee, in a less than critical appraisal of food irradiation, "may not sufficiently take account of, still less exclude, possible long-term medical effects on the population, given that irradiated products have been available for a relatively short period of time...more scientific data is required...and a full-scale study should be undertaken...before the process can be confidently accepted."

Thus the BMA opposes irradiated food. Based on the same facts, what does the American Medical Association (AMA) have to say?

Contrary to the BMA and the serious questions it has raised, the AMA seems to give its blessing to food-zapping: "Many years of international experience," says the AMA, "have demonstrated that food irradiated at levels of up to 1 million rads (40 million chest X-rays) are safe to eat...Food irradiation produces no significant reduction in the nutritional quality of food." Commenting on a food industry sponsored bill to reclassify food irradiation as a "process," not a food additive, thereby weakening FDA's regulatory authority over irradiation by reducing safety testing requirements, the AMA had this to say: "In our view formal official reclassification of food irradiation is important in terms of public acceptance of the fact that food irradiation is a safe process, not a potentially hazardous food additive."

Thus our AMA, as opposed to the more truly conservative BMA, throws caution to the wind and ignores the evidence. Pork, irradiated at 300,000 rads, much less than the level the AMA "believes" to be safe and nutritious, has been found in U.S. Department of Agriculture studies to lose 46.5 percent of the essential nutrient, thiamine.

Rejecting the false sense of security that the AMA shares with its friends in the food-irradiation industry, American consumers are more in tune with the cautious BMA.

According to a recent article in *The New York Times,* many elements of the American food industry are joining the ranks of the cautious. Ellen Green, a spokeswoman for the National Food Processors Association, has said that "The food industry is a little bit spooked by the adverse reaction they've been receiving from consumer groups." In New York, spokespersons for Pathmark, Sloan's, A&P and Food Emporium chains said the stores would probably not sell irradiated produce because of the potential for bad consumer reaction.

Although producers of fruits and vegetables were given the green light by the FDA to use irradiation a year ago, there is no detectable rush to use it, partly because it strips away the color of most green vegetables. As far as the extension of fruit and vegetable shelf life, many producers know that storage in a temperature- and humidity-controlled atmosphere can accomplish the same objective.

Last year, FDA commissioned a study to find out why there was so much consumer opposition to food irradiation. The study, by HIS Associates in Washington, D.C., found that the information FDA was trying to convey to the public was "clear, consistent, constrained." The constraint, said HIS, "was good scientific practice," a reason being:

"Scientific uncertainty - with over 40 years of research and experience, much is known about food irradiation. But gaps remain in that knowledge. In order to make a regulatory decision about irradiation of food, FDA had to make a scientific judgment, in spite of the remaining uncertainty about the effects of irradiation at different dose levels."

We fully agree that there is "uncertainty" about the safety of irradiation. We thereby believe FDA made the wrong decision in approving its use and that the AMA has much to learn from the BMA (and the American public).

technologies other than irradiation that appear equally effective but lack the drawbacks of irradiation. One such treatment is the use of microwaves to heat the grain under vacuum conditions. This technique, developed by the McDonnell-Douglas Corporation, the Aeroglide Corporation, the U.S. Department of Energy, and USDA, is ready for commercialization. Heat treatment by infrared also appears to be feasible and effective.

Irradiation has also been promoted as a means to reduce spoilage of fruits and vegetables, especially produce destined for under-developed countries. In a speech to the New York Society of Security Analysts, Dr. Martin Welt, Chairman of the Board of Radiation Technology, made the following statement:

"Consider for a moment, the advantages of the latter process [irradiation] for Third World nations lacking adequate distribution networks or refrigeration systems. Remember one billion Chinese as a large potential market where household refrigeration is nonexistent."

However, large scale testing under real-life conditions has demonstrated that irradiation is not very well-suited to extending shelf-life. In 1971, scientists at the University of California at Davis conducted a comprehensive study of the feasibility of using irradiation to extend the shelf-life of 20 different commodities. In their paper they stated:

"Without exception in all commodities studied, irradiation increased susceptibility to transit injury. If stationary tests alone had been used with boysenberries, nectarines, peaches, raspberries, and tomatoes, irradiation might have been judged effective. In actual and simulated transit, mechanical injury of all these fruits was intolerably high in irradiated specimens."

The authors concluded by saying:

"[T]he diversity of species studied and the negative results obtained indicate that irradiation holds little promise for perishable commodities...Occasionally, irradiation is suggested as a promising substitute for refrigeration. This hope should be dispelled with respect to controlling rot or senescence in fruits and vegetables."

These studies found that radiation caused significant softening of the flesh of the fruits and vegetables, leading to considerable transport injury.

In a recent conversation, one of the University of California researchers, Dr. Noel Sommer, reaffirmed his belief that irradiation held little promise as a means to extend the shelf-life of fruits and vegetables. He stated that the benefits of irradiation at the doses that would be permitted by FDA were "very minimal."

While elimination of Trichinella spiralis from pork sounds like a major accomplishment, proponents of irradiation for this purpose fail to note that trichinosis from commercially prepared pork products simply is not a major problem in the United States. In 1984, the Centers for Disease Control reported less than 50 cases of trichinosis from pork products.

In addition, there are alternative methods to assure that pork products are not contaminated with this parasite that are easier and less expensive than irradiation. According to Dave Meisinger of the National Pork Producers Council, pork carcasses can be individually tested for the presence of Trichinella spiralis for about 15 cents each using a method known as Enzyme-Linked Immuno-Sorbed Assay (ELISA). The Department of Energy has estimated that irradiation would cost between 27 and 95 cents per hog. The other benefit of ELISA is that the infected hog can be traced back to a particular farm where the farmer can be made to clean up his operation. Besides, simply cooking pork thoroughly will destroy the parasite.

Because the need for irradiation of pork is doubtful, there have been suggestions that its use is mainly being considered for pork intended for export. Since other countries do not have pork that is as clean as that raised in the U.S., being able to certify pork as "trichina free" is considered a major marketing advantage overseas. Irradiation, its proponents hope, can accomplish this aim. Unfortunately for those proponents, however, no foreign government has yet given its approval for importation of irradiated pork. In September of 1985, USDA asked U.S. embassies in 40 different countries to ask the host government if they would accept shipment of irradiated pork. To date, none of these governments has said it would accept it. In addition, West Germany and Sweden have indicated that they will not accept irradiated pork. Without the approval of the importing country, USDA will not allow these shipments.

Possible Nutrition Loss From Irradiation

While the possible loss of nutrients in food subjected to irradiation needs to be studied more extensively, there is some preliminary data that suggests that irradiating pork may lead to a substantial loss of thiamine. An initial study conducted at the USDA's Eastern Regional Research Center in Philadelphia found that pork irradiated at 100 krads (the amount permitted by FDA) suffered a 20 percent loss of thiamine. The thiamine level in pork irradiated at 500 krads was reduced by 54 percent. Pork consumption accounts for 14 percent of America's thiamine intake.

HRG Actions to Date

Whenever FDA finally approves the use of a food additive, including the use of irradiation, the public has thirty days in which to file objections with the FDA. The FDA is then required to hold a hearing "as promptly as possible" to assess the merits of these objections.

After the July 22, 1985 FDA approval of pork irradiation, HRG filed objections and requested a hearing. As of February 3, 1986, the FDA ha[d] not granted the required hearing. Until this hearing takes place, and a decision is issued, we are precluded from challenging this approval in court. HRG also plans to file objections to the FDA approval of irradiation of fruits and vegetables as soon as the approval is finally announced in the Federal Register.

Update: The FDA Gives a Green Light to Food Irradiation

On April 15, 1986 the Food and Drug Administration (FDA) approved the use of irradiation treatment for fruits and vegetables, herbs and spices, white potatoes, and wheat and wheat flour. (Irradiation is already allowed for the treatment of pork.) Under the new regulations, radiation doses of up to one kiloGray or 100,000 rads - a dose of radiation equivalent to more than 30 million chest X-rays - can be used to treat fruits and vegetables, and up to 30 kiloGray can be used to disinfect spices and herbs. According to the Produce Marketing Association (a trade group), papayas and mangoes are likely to be the first irradiated products that consumers can expect to see on store shelves. In fact, irradiated mangoes have already appeared in Florida markets.

Labeling Irradiated Foods

Under the new FDA regulation, irradiated products must contain labels stating that the product has been "treated with radiation" or "treated by irradiation." This is better than the FDA's original proposal to require irradiated foods to be obscurely labeled as "picowaved." However, the regulation is far from perfect: foods that contain irradiated ingredients, but have not themsleves been irradiated, do not have to be labeled. For example, irradiated foods that are used as ingredients in processed foods (such as canned soups or frozen pizza) and foods served in restaurants, dormitories, hospitals, prisons, and the military, do not have to be labeled. Two years from now, irradiated foods will only have to be labeled with a misleading flower-like symbol, leaving consumers even more in the dark about what they're eating.

The Case Against Food Irradiation

The FDA regulations were issued over the objections of Public Citizen Health Research Group (HRG) and many other organizations and individuals concerned about potential safety and health problems of food irradiation. HRG has argued that the safety of food irradiation has not been proven, and that studies have found kidney and testicular damage, as well as chromosomal damage, in animals fed irradiated food. Also, the possibility of a risk of cancer from consuming irradiated food cannot be ruled out since studies have shown an increase in tumors in mice and rats fed irradiated food.

Moreover, recent charges brought against two facilities that use radiation to sterilize medical equipment and food have fortified our fears that food irradiation will pose environmental threats to workers and surrounding communities:

• On June 23, 1986, the Nuclear Regulatory Commission (NRC) closed and requested a Justice Department criminal investigation of a Radiation Technology, Inc. plant in Rockaway, NJ for violating NRC safety requirements and lying to NRC officials. Company personnel were charged with bypassing safety interlocks (systems designed to prevent workers from entering the irradiation chamber and receiving high doses of radiation) and "willfully providing false information to the NRC staff." The NRC concluded that it "lacked reasonable assurance that the facility would be operated in a manner that assured that the health and safety of the public, including licensee employees, would be protected." In 1977, an employee was exposed to a near-fatal dose of radiation as a result of the company's safety violations. Martin Welt, the president of the company and a major proponent of irradiation, could face criminal prosecution and a possible jail sentence for his company's conduct.

• On July 3, 1986, the Department of Agriculture notified Welt that it was withholding approval of the company's plan to irradiate fresh pork pending an investigation into "false information" that had been submitted by the company on its quality control program.

• On July 9, 1986, the NJ Department of Environmental Protection notified the Radiation Technology plant that it is conducting a $600,000 study to determine the extent of toxic contamination in the groundwater surrounding the plant. Methylene chloride contamination at the Radiation Technology site has caused the EPA to list the plant as a hazardous waste site. The state plans to sue the company to recover expenses.

• The NRC has also obtained a criminal indictment against International Nutronics Inc. and two company officials for attempting to cover up a spill of radioactive water from its Dover, NJ plant by flushing it down a bathroom drain leading into the Dover sewer system. The cleanup will cost more than $2 million. Company officials will now have to stand trial for their actions.

SECTION 2
What is the Federal Government Doing?

There are long-standing laws to police the safety of the food supply – reflecting one of the most basic domestic functions of government. The implementation of these laws for the public health reflect a tension between the companies who are expected to comply with them and the consumers who are supposed to benefit from them. Guess who prevails much of the time? The powerful food lobbies. But this dominance is one of default due to the inadequate organization of consumers–a vacuum that recently is being filled by several Washington-based groups with grass-roots supporters. This section provides a specific controversy by controversy status report compiled by Marian Burros of The New York Times along with a General Accounting Office summary of inadequate regulatory authority and weak enforcement by the Food and Drug Administration.

Other materials depict the serious weaknesses in federal pesticide control programs and the current effort in Congress to strengthen these laws and, for once, make them enforceable. An exciting new frontier, which is rediscovering much old knowledge, for "getting off the pesticide treadmill" is the topic treated in *Technology Review* by Michael Dover. When you read this article about more benign and effective pest controls, can you but wonder why commercial interests got agriculture on this toxic treadmill in the first place? Then an article titled "Everything Doesn't Cause Cancer: But how can we tell which chemicals cause cancer and which ones don't?" by the U.S. Public Health Service is included to counteract any overdrawn reader reactions.

Principal Laws Governing Food Safety

Listed below are the principal federal laws governing the safety of the food supply. Under these laws there are regulations and standards that are much more detailed. The federal agencies, your member of Congress, or some local law libraries should be able to provide access to these laws and regulations or, within reason, with copies of the sections you want. It should be noted that what the laws say should be done is very often quite far from what the agency or department is able or willing to implement. Also note that there are state and local laws governing food safety and inspection right down to the retail level of restaurants and food stores.

A number of these federal agencies could use still more legislative authority. This is especially true of the FDA. Under current law, food processors have no obligation to notify FDA when products subject to their control have become adulterated, or when they recall products because they believe the products may be in violation of the law. Food processors are not required to permit FDA to inspect food processing records that may bear on whether the products are adulterated or mislabeled. FDA does not have subpoena power, the authority to compel the production of documents or the testimony of witnesses, and it has no embargo or detainment authority. Many weeks may pass between the time of an inspection of a food processing facility and the filing in court of papers to effect a seizure or obtain an injunction concerning violative products; in the meantime, the firm holding the violative products may ship them into the stream of commerce.

1. The Federal Food, Drug and Cosmetic Act, enacted in 1938 and amended in 1958 and 1960, was passed to protect the public from potential cancer-causing chemical food additives. Under the law all food dyes are supposed to be proven safe before they are used. Also under the law, the FDA has the authority to preserve the safety of the nation's foods by prosecution, seizure, injunction and recalls. The FDA also regulates the use of animal drugs and intentional additives in food and sets tolerances or action levels for contaminants that are unavoidably present in food or feeds.

2. The pesticide tolerance setting functions, governed by the Federal Food, Drug and Cosmetic Act, were transferred in 1970 to the newly created Environmental Protection Agency. The pesticide registration system, administered under the Federal Insecticide, Fungicide, and Rodenticide Act of 1947, currently being reviewed, was moved from the U.S. Department of Agriculture to the Environmental Protection

Agency in the same year.

3. The Federal Meat Inspection Act (1906), the Poultry Products Inspection Act (1957), and the Wholesome Meat Act (1967) require that livestock and poultry slaughtered at plants that do business in interstate or foreign commerce be federally inspected. Federal inspection is also required at slaughter plants that do intrastate business in states not having their own inspection program. These laws require that federally inspected meat and poultry plants operate in a sanitary manner and that the products they sell be wholesome, unadulterated, and properly labeled. Plant managers are primarily responsible for meeting these requirements. The Food Safety and Inspection Service of the U.S. Department of Agriculture is responsible for meat and poultry inspection activities. The FSIS devises ways to ensure that harmful chemical residues are not present in meat, poultry, and egg products.

4. The Agricultural Marketing Act of 1946 provides for the development and promulgation of grade standards, and the inspection and certification of transportation and processing facilities for fish and shellfish and all processed fish products. The authority for these regulations was transferred to the U.S. Department of Commerce in 1971.

5. The Safe Drinking Water Act of 1974, as amended, requires the Environmental Protection Agency to establish federal standards for drinking water, protect underground sources of drinking water, and establish a joint federal-state system for assuring compliance with the resultant regulations.

The Safe Drinking Water Act (SDWA) of 1986, the amended version of the act originally passed in 1974, included much stricter requirements on standard-setting for drinking water contaminants. Legislative deadlines were set for the EPA in order to avoid the slow pace of standard setting that characterized the previous 12 years and that Sen. Durenberger of Minnesota called "miserable, discouraging, disturbing..."

The Act required the EPA to set nine standards by June 1987, (the EPA set eight standards by that date), 40 standards in total by June 1988, and 83 standards in total by June, 1989. By January 1988, the EPA is also required to establish a list of substances known or anticipated to occur in drinking water, and all community water systems will be required to monitor at least once every five years for these substances.

The amendments also require the EPA to specify criteria for filtration of surface water supplies, and disinfection of groundwater supplies; to protect sole-source aquifers supplying drinking water, and to propose treatment as a solution to contamination problems. The use of solder and flux containing more than 0.2 percent lead is prohibited, and the use of pipes containing more than 8 percent lead in new installations and in repairs are banned under the amendments. Public notification requirements for contamination problems were also upgraded in the amendments. The Act as amended applies to all public water systems, that is, all systems publicly or privately (investor) owned that serve an average of at least 25 people at least 60 days of the year. The SDWA does not apply to private wells.

U.S. Food Regulation:
Tales From a Twilight Zone
BY MARIAN BURROS
NEW YORK TIMES, JUNE 10, 1987, P. C1.

Whatever happened to cyclamates, the artificial sweeteners we learned to love and then, with talk of cancer, leave? How about apples sprayed with Alar, candy dyed with Red No. 3 food coloring, irradiated foods and sodium nitrite in processed meats? Has the Federal Government decided whether these substances and treatments are hazardous to human health? If they are hazardous, why are they still in use?

Periodically, questions about food safety capture public attention, then just as quickly disappear. But even as the spotlight shifts, these issues are followed by contending interest groups, including food manufacturers and consumer and environmental groups. The resulting regulatory delays have lasted sometimes for decades, as the accompanying list of case histories indicates.

Present and former officials of the Food and Drug Administration and the Agriculture Department agree on their role in this regulatory drama.

"There is no question that the F.D.A.'s mission is to protect consumers," said Dr. Donald Kennedy, the agency's commissioner during the Carter Administration and now president of Stanford University.

But officials differ over the approach. "Whether the right approach is a heavy regulatory one or basically to encourage innovation within the industry," Dr. Kennedy said, "I think you will find that Republican administrations tend to employ the latter and Democratic administrations tend to employ the former."

Since the late 1970's, however, both Republican and Democratic administrations have cut back on agency budgets.

"The budget has not grown to keep pace with the growth of responsibilities," Dr. Kennedy said. "But we have always tried to buy that particular service on the cheap."

Couple the shortage of money with the inherent slowness of bureaucracies and the result is a nether world of unfinished business.

"If the regulatory system works at all, it works slowly," said Dr. Michael Jacobson, executive director of the Center for Science in the Public Interest, a Washington consumer group.

Dr. Sidney Wolfe, director of the Public Citizen Health Research Group and a longtime critic of Federal regulatory policies, said enforcement of laws and regulations too often depends on "pressures from the food industry trickling down through the White House, the Office of Management and Budget, the Secretary of Health and Human Services and the Secretary of Agriculture."

"Americans are not being adequately protected from cancer-causing or other dangerous substances in the food supply," Dr. Wolfe said.

Members of the food industry generally disagree.

"I can guarantee you that both consumer groups and food companies say they are arguing on behalf of the consumer's health," said Dennis Phelan, director of legislative affairs for the National Food Processors Association, an industry group.

"A lot of consumer groups start from the position that there should not be any chemicals in food that are potentially harmful, regardless of quantity," he said. "The food industry's position is that you should make a scientific assessment of chemicals that may be present and if there is not a health risk then there should not be a problem."

The current commissioner of the F.D.A., Dr. Frank E. Young, said there are reasons for the slow pace: "The complexity of the problems, the demands placed on F.D.A. with the vast array of emergencies, conflicting priorities and being sure we have our scientific ducks in a row."

Dr. Jere Goyan, commissioner during the Carter Administration, said the agency is "very good on immediate life-threatening matters, basically all those things everyone's against, such as botulism." The areas in which the agency has trouble, he said, "are where there's a sponsor who wants to use a material that others think might be carcinogenic."

The result is a hodgepodge of regulations that have been proposed but never made final; that have been under study for 5, 10 or 25 years; that have been in litigation for even longer. It is a history of legislative and regulatory delay that serves neither the public nor the food industry. ◆

AFLATOXIN

Naturally occurring toxin, produced by molds, in grains and peanuts.

Early 60's - Aflatoxin discovered.

1965 - F.D.A. sets first permissible level, or action level.

1969 - First study shows toxin is a carcinogen.

1980 - Public Citizen, a consumer group, sues to require public hearings when action levels are being set.

1985 - Public Citizen wins suit.

1986 - Supreme Court reverses decision; sends case down to be retried on other grounds.

Status: Public Citizen wins again; products containing aflatoxin will be prohibited unless F.D.A. demonstrates safety of specified levels. Decision affects other unavoidable contaminants, like mercury. F.D.A. might appeal.

ALCOHOL LABELING

Alcohol is the only category of food for which ingredient labeling is not mandatory. Many people are allergic to ingredients, like sulfites, in alcohol.

1972 - Center for Science in the Public Interest, a consumer group, petitions Federal agencies to require labeling.

1980 - Treasury Department under the Carter administration issues rule requiring partial ingredient labeling, or an address on label where ingredient information is available.

1981 - Reagan Administration rescinds rule.

Status: Treasury Department agrees on labels for potentially unsafe ingredients; saccharin and Yellow No. 5 food coloring must be listed. Some bottles must list sulfites beginning this month; all others have until January (1988).

ANTIBIOTICS

To stimulate cattle growth and prevent disease, producers use penicillin and tetracycline, antibiotics used in humans. Bacteria in animals can become resistant to the antibiotics; humans infected with antibiotic-resistant bacteria may not respond to treatment with those antibiotics.

1972 - F.D.A. investigates health effects.

1977 - F.D.A. proposes ban on all uses of penicillin and most uses of tetracyline in animal feed. Congress says further studies are needed.

1980 - National Academy of Sciences can reach no conclusion on health risks, but recommends using other growth stimulants.

1984 - Centers for Disease Control document outbreak of antibiotic-resistant salmonella. Natural Resources Defense Council, an environmental group, petitions Department of Health and Human Services for ban.

1985 - Petition denied.

1987 - Centers for Disease Control document another outbreak of antibiotic- resistant salmonella.

Status: F.D.A. says studies are needed. Some beef producers are trying antibiotics not used in humans.

ASPARTAME

Sugar substitute marketed as NutraSweet and Equal.

1965 - Aspartame discovered.

1970 - It is found that aspartic acid, an aspartame component, produces brain lesions in test mice.

1974 - F.D.A. approves use; agrees to inquiry into sweetener's safety.

1980 - Inquiry board advises more testing; would forbid marketing.

1981 - F.D.A. overrules board; approves aspartame in dry foods.

1983 - F.D.A. approves aspartame in carbonated beverages. New studies suggest it might be serious health hazard; F.D.A. alerted. Reports of side effects, such as brain seizures, increase.

1986 - Community Nutrition Institute, a Washington consumer group, petitions F.D.A. to ban aspartame. Petition rejected. On appeal in Circuit Court of Appeals in Washington D.C.

Status: Studies continuing. Aspartame replacing saccharin in diet beverages.

BHA, BHT

Butylated hydroxyanisole and butylated hydroxytoluene are anti-oxidants used in breakfast cereals, vegetable oils, processed meats, snacks and spices to extend shelf life and prevent rancidity. Usually listed on labels. Some evidence they cause cancer.

1977 - Center for Science in the Public Interest petitions F.D.A. to ban BHT. F.D.A. proposes restricting BHT use. Proposal never becomes final.

1982 - Japanese researchers find BHA causes cancer in test animals.

1986 - World Health Organization's International Agency for Research on Cancer finds "sufficient" evidence that BHA is a carcinogen, "limited" evidence that BHT is a carcinogen.

Status: Both still in use; safer alternatives, like Vitamins C and E, also being used.

CYCLAMATES

Artificial sweeteners listed in 1958 among additives Generally Recognized as Safe (GRAS).

1951 - F.D.A. ignores first studies that show cyclamates are carcinogenic.

1969 - New pathologist at F.D.A. shows 1951 study to National Academy of Sciences, adding to evidence for banning cyclamates.

1970 - Cyclamates banned. Manufacturer, Abbott Laboratories, asks for reapproval.

1980 - Requests rejected.

1985 - National Academy wants more studies.

Status: F.D.A. reviewing data.

FISH INSPECTION

Procedure, now voluntary, that some want mandated from catch to point of sale.

1967, '68, '69, '71, '74 - Congressional hearings on legislation requiring inspection.

1984 - Mandatory inspection recommended by General Accounting Office, Congressional investigative arm. Bill mandating inspection dies.

1986 - Public Voice for Food and Health Policy, a Washington consumer group, releases report documenting health risks of not having mandatory inspection.

1987 - Bill mandating inspection is introduced. Senate Agriculture Committee holds hearings. Dairy, Livestock and Poultry Subcommittee of House Agriculture Committee expected to hold hearings this summer.

Status: Only 15 percent of fish and seafood is voluntarily inspected.

ARTIFICIAL FOOD COLORS

Colors made from petrochemicals are used to enhance food appeal. Seven are in use: Blue Nos. 1, 2; Green 3; Reds 3, 40; Yellows 5, 6. Artificial colors must be listed on food labels except dairy products like butter, cheese, ice cream.

1960 - Congress gives industry 2 1/2 years to submit test results on safety to F.D.A. Deadline has been extended more than 20 times since then.

1976 - F.D.A. gives industry four years to complete tests.

1977, 1981 - Public Citizen Health Research Group sues F.D.A. over delays; loses both suits.

1983-85 - Test results on remaining colors submitted. Red No. 3 is judged a carcinogen.

1983 - F.D.A. proposes permanent approval for Blue No. 2, despite conflicting evidence of carcinogenicity.

Public Citizen seeks ban.

1984 - F.D.A. asks Department of Health and Human Services for approval to ban Red No. 3; approval denied.

1985 - Public Citizen sues F.D.A. again to complete testing on Yellows Nos. 5, 6 and to ban Red 3; loses. Case on appeal.

1987 - F.D.A. denies requested ban on Blue No. 2. Public Citizen sues to reverse decision.

Status: Several colors in limbo. Some manufacturers turn to natural colors. From 1950 to 1984, per capita consumption of artificial colors triples.

MEAT AND POULTRY INSPECTION

Agriculture Department inspectors in slaughterhouses check for microbiological contamination; random testing for chemical residues.

1906 - Federal Meat Inspection Act passes, requiring inspection of all meat sold in interstate and foreign commerce.

1957 - Poultry Products Inspection Act passes (similar to 1906 act).

1967, '68 - Wholesome Meat Inspection Act and Wholesome Poultry Products Inspection Act extend inspection to intrastate sales.

1970's - Agriculture Department changes from inspecting every bird to sampling method.

1985, 1987 - National Academy of Sciences judges meat and poultry inspection procedures to be inadequate; recommends new poultry inspection system.

Status: Bribery and extortion charges against inspectors and plant owners are a small but continuing problem, raising questions about inspection standards. Increases in food-borne illnesses from meat and poultry are recorded.

DAMINOZIDE

Chemical, trade name Alar, used on crops for uniform ripening.

1968 - Registered for use on apples.

1973 - Study indicates possible carcinogenicity in test animals.

1980 - Environmental Protection Agency reviews daminozide for possible withdrawal of registration.

1984 - E.P.A. says daminozide is a carcinogen.

1985 - E.P.A. issues draft report on proposed ban, but agency advisory panel says data are inadequate; calls for more testing.

1986 - Ralph Nader asks supermarkets to stop buying apples treated with daminozide. Consumer and environmental groups, New York State and Maine petition E.P.A. for ban.

1987 - Request denied, but agency lowers permissible levels.

Status: Consumer groups sued E.P.A. Case pending. Many supermarkets and manufacturers of apple products refuse apples treated with daminozide.

SACCHARINE

Artificial sweetener discovered in 1900's. Used by diabetics

1955-65 - Begin use in diet foods.

1960- 70's - Studies find saccharin causes bladder cancer in test animals; epidemiological evidence of saccharin as a human carcinogen.

1977 - F.D.A. announces ban; Congress institutes 18-month moratorium on ban.

1980, '81, '83, '85 - Further moratoriums voted.

1987 - Senate passes five-year moratorium.

Status: Products containing saccharin must carry warning labels. Use has dropped since introduction of aspartame.

URETHANE

Result of manufacturing techniques, a carcinogen in some wines, distilled spirits, imported fruit bran-

---

The Rule on Additives

THE BASIS FOR THE MAJORITY OF Federal food regulations is the Federal Food, Drug and Cosmetic Act of 1938. But a 1958 amendment, the Delaney Clause, has caused the greatest controversy. The amendment prohibits the use of food additives that cause cancer in animals or humans. An amendment on food coloring was added in 1960.

Since 1977, when the Food and Drug Administration invoked the Delaney Clause in its unsuccessful effort to ban saccharin, Congress has made many attempts to repeal or modify the clause.

In 1981, a proposal by Senator Orrin G. Hatch, Republican of Utah, to repeal the Delaney Clause died in committee.

In 1983, Senator Hatch introduced another bill to repeal; it also died.

In 1985, the F.D.A. instituted a more liberal interpretation of the clause, under which the agency can allow known carcinogens in food if it determines the risk to humans is negligible. Public Citizen, a consumer group, has argued in court against this interpretation, saying it would gut the Delaney Clause. A decision is pending.

dies. In other alcoholic beverages, presence varies. Not found in beer.

1985 - Canada discovers urethane as a contaminant, establishes tolerance levels.

1986 - Center for Science in the Public Interest petitions F.D.A. to alert public and publicize which brands containing urethane.

Status: Center suggests F.D.A. recall brands with high levels. Agency refuses; under review.

METHYLENE CHLORIDE

Chemical solvent used to extract caffeine from coffee. Until mid-70's, trichloroethylene, another chemical solvent in the same family, dominated the market.

1975 - Public Citizen petitions F.D.A. to ban trichloroethylene, which causes cancer in test animals. F.D.A. takes no action, but companies withdraw it and substitute methylene chloride.

1985 - F.D.A proposes ban on methylene chloride in cosmetics (where it is used as propellant or solvent) because it causes cancer in test animals. Agency will not ban it in coffee, saying small residue poses negligible risk.

Status: Public Citizen filed lawsuit; arguments scheduled for September in Washington. Many coffee producers have switched from methylene chloride to other processes, but labeling is not required.

SODIUM NITRITE

Used to preserve, color and flavor processed meats.

1969 - Scientists find sodium nitrite reacts with amines (chemicals in foods, saliva, drugs) to form nitrosamines (potent carcinogens).

1972 - Consumer groups petition Agriculture Department to ban unnecessary use of nitrites and establish a study committee. Committee calls for additional research.

1978 - Agriculture Department and F.D.A. prepare to ban nitrites based on M.I.T. study showing it is carcinogen. F.D.A. review suggests not so.

Status: Nitrites still used but in smaller amounts in processed meats; meats available without nitrites are labeled as such.

RAW MILK

Milk that is not pasteurized to protect against various food-borne illnesses.

1972 - F.D.A. begins regulatory process to prevent sale of raw milk. Industry obtains a stay.

1984 - Public Citizen petitions F.D.A. to ban interstate and intrastate sales. F.D.A. delays action.

1985 - Public Citizen asks court to force F.D.A. decision and wins. F.D.A. denies petition. Public Citizen files complaint in Federal District Court in Washington, seeking review of F.D.A. action.

1986 - Court orders F.D.A. to ban interstate sales.

Status: Awaiting notice of F.D.A. proposal to ban. Intrastate sales not affected. Nation's largest seller of raw-milk products, Alta Dena, can continue sales in California, which account for 95 percent of raw milk sales in the country; 20 to 30 states ban sales of raw milk within their borders.

SULFITES

Preservatives used in fruits, vegetables and wines to prevent bacterial growth, discoloration; masks spoilage.

1982 - F.D.A. proposes adding to Generally Recognized as Safe list. Center for Science in the Public Interest discovers studies showing sulfites can cause asthmatic attacks, urges F.D.A. ban; 1982-1986, six recorded sulfite-related deaths.

1985 - Congressional hearings triggered by death of 10-year-old Oregon girl.

1986 - F.D.A. bans sulfites on most fresh fruits, vegetables; enforces labeling laws more strictly.

Status: This month, Treasury Department required declaration of sulfite content in alcoholic beverages. Sulfites in dried fruits must be labeled; regulation on potatoes is awaited.

Raw Milk Ban

HEALTH LETTER, VOLUME III, #2, MARCH 1987

Ruling in a lawsuit in 1985 by Public Citizen Health Research Group Director Sidney M. Wolfe, M.D. and the American Public Health Association, U.S. Federal District Court Judge Norma Johnson said on December 31, 1986, "There is no longer any question of fact as to whether the consumption of raw milk is unsafe." She ordered the Food and Drug Administration (FDA) to ban interstate sales of raw milk and raw milk products. Although the ban does not affect intrastate sales, 25 states have already banned such sales, and many more are expected to follow with their own bans in the wake of the court order. The last state to ban raw milk is likely to be California, home of the powerful Alta-Dena Dairy in Los Angeles – the only major U.S. producer of raw milk still selling this dangerous food product. Through Alta-Dena's efforts, the FDA has been previously stopped in its tracks three times as it was on the verge of banning interstate raw milk sales.

Since 1980 alone, more than 1,000 Americans have become ill because of consumption of raw milk; more than 20 have died. Action: Ask your state senator or representative to co-sponsor legislation to ban raw milk sales, especially if you live in California (or New York, New Jersey, Connecticut, or other states soft on raw milk). The victims are often infants, the very old, and patients with immune deficiency such as cancer patients getting chemotherapy. They are usually victims because of unproven Alta-Dena claims that raw milk will improve health. ◆

FDA Admits Laxity in Testing Food Imports

BY MICHAEL WEISSKOPF

THE WASHINGTON POST MAY 1, 1987

Five years after the cancer-causing pesticide heptachlor was banned in the United States, the Philippines was annually exporting hundreds of thousands of tons of heptachlor-treated pineapple to this country.

None of it was stopped – or even tested – by the Food and Drug Administration.

The pineapples were cited yesterday at a House Energy and Commerce subcommittee hearing as evidence of the FDA's failure to stop imports of food adulterated with pesticides, including cancer-causing chemicals banned in this country.

The FDA commissioner, Dr. Frank E. Young, acknowledged that inspection was lax, but declared that "we do not find that the food supply is contaminated" by pesticides.

Testimony showed that little of the 45 billion pounds of food shipped annually into the United States is sampled for pesticides, that 64 percent of the imported food products sampled from 1982 to 1985 contained pesticide residues, and that half of the imports found to be adulterated moved to market before the FDA had completed its analysis and ordered them detained.

Of the food imports, valued at $20 billion last year, Assistant Comptroller General J. Dexter Peach said, "Not only is there inadequate coverage of certain pesticides, but some foods coming from various countries are not being tested for any pesticides at all."

The subcommittee was told of a pesticide group known as EBDCs, the chemicals most widely used worldwide against fungus in fruits and vegetables.

Although a byproduct of EBDCs is known to be a probable cause of human cancer, Peach said the FDA did not test any imported food for the pesticide from October 1978 through last March. At the same time, according to the subcommittee, imports of such EBDC-treated products as tomatoes and orange juice concentrate increased dramatically.

"That record is unacceptable," said Rep. Ron Wyden (D-Ore.), directing his criticism at Young.

"I agree, and that's why we're trying to change it," Young replied, pledging increased enforcement of pesticide limits on imported foods.

A General Accounting Office analysis presented by Peach showed that 6 percent of imported food samples surveyed by the FDA between 1979 and 1985 violated federal pesticide limits. There was evidence of violations by domestic producers as well; 2.9 percent of the U.S. food had excessive levels.

But yesterday's hearing focused on pesticide residues in imports, which account for one-sixth of U.S. food consumption, including one-fourth of the fruit.

Rep. Gerry Sikorski (D-Minn.) was critical of the FDA sampling of imported bananas, of which only two were tested from 1979 to 1987 despite wide foreign use of the pesticide benomyl, suspected of causing cancer, birth defects and genetic mutations.

"The FDA can no longer be bashful about hazardous food imports," he said. "The time has arrived to subject the Chiquita banana lady to a serious strip search." ◆

Legislative Changes and Administrative Improvements Should be Considered for FDA to Better Protect the Public from Adulterated Food Products

A Summary of the Comptroller General's Report To The Chairman, Subcommittee On Oversight and Investigations, House Committee On Energy and Commerce
SEPTEMBER 26, 1987

Digest:

The Food and Drug Administration (FDA) is responsible for assuring that food offered to the public is pure, safe, wholesome, and properly labeled. Food found to be adulterated, mislabeled, or potentially harmful (adulterated food products) may be voluntarily recalled by food firms under FDA's direction or seized through legal action.... Firms producing adulterated products and officials of such firms may be prosecuted. . . .

FDA Must Rely On Others [Federal Agencies] To Detain Adulterated Foods:

Although the Federal Food, Drug, and Cosmetic Act (FD&C Act) authorizes FDA to seize adulterated food products, it does not give FDA authority to detain – prohibit movement or marketing – such products while processing seizure actions through a U.S. attorney's office and the courts. Once the court approves the seizure, a U.S. marshal is ordered to seize the product. On the average, this process takes about 65 days....

FDA could prevent greater amounts of adulterated products from getting on the market if it had legislative authority to detain adulterated food products while it obtains a seizure notice....

FDA Could Accelerate Its [Aldulterated Food Products] Seizure Approval Process:

GAO's review of 202 seizures showed that FDA's review process took 41 days on the average, ranging from five to 206 days. On the average, an additional 24 days, ranging from one to 150 days, was needed to process the seizure through the U.S. attorney's office and the courts. . . .

[M]ore timely actions could help keep more adulterated products from reaching the market.

FDA Needs Authority To Review Manufacturers' Records Once Adulterated Products Are Identified...

FDA's ability to remove adulterated products from the market is hampered because it has no legislative authority to review a manufacturer's production and distribution records....

FDA Needs To Improve Its Monitoring Of Recalled Food:

FDA has not developed guidelines for its districts to use in verifying the reconditioning or destruction of recalled food....

Problems can arise because of inadequate FDA verification. During recent recalls, involving significant health hazards, products were later sold without being properly reconditioned. For example, in one case, because the product was sold to retail outlets without being properly reconditioned, it had to be recalled a second time....

Maximum Fine For Violating The [F,D&C] Act Should Be Increased:

In 1938, the Congress believed that fines higher than those in existence were needed to bring about compliance with the FD&C Act and raised the fines to the current maximum level. However, the amounts established have never been adjusted and have been eroded by inflation....

Fines assessed are small, considering that each of the firms had estimated annual sales that exceeded several hundred thousands of dollars.

Recommendations To The Secretary Of Health And Human Services:

GAO recommends that the Secretary direct the Commissioner of FDA to:

• Initiate steps to improve the timeliness of seizure actions by identifying more routine seizure cases involving filth and economic adulteration that could be referred by district directors directly to U.S. attorneys after concurrence by FDA's General Counsel.

• Develop guidelines specifying required verification procedures to ensure that the destruction or reconditioning of recalled foods is adequately verified directly by FDA or through some alternative means, such as appropriate state or local officials.

Matters For Consideration By The Congress:

The Congress should consider amending the FD&C Act to provide FDA authority to detain food products suspected of being adulterated and to review manufacturers' production and distribution records of adulterated products. Also, the Congress should consider increasing the maximum fine associated with criminal prosecutions for persons or firms convicted of violating provisions of the act....

Pesticides in Domestic Foods

Executive Summary, General Accounting Office

OCTOBER 1986

Purpose:

Pesticides are used extensively in American agriculture and residues of these pesticides on food need to be closely monitored to protect the public from harmful effects. The Food and Drug Administration (FDA) is responsible for monitoring the domestic food supply to identify food with illegal residues and to remove it from the market. Illegal pesticide residues are those that are not allowed to be present on food or are present in greater concentrations than those authorized by the Environmental Protection Agency. Such food is adulterated and cannot legally be marketed in interstate commerce.

The Chairman and Ranking Minority Member, Subcommittee on Toxic Substances and Environmental Oversight, Senate Committee on Environment and Public Works, asked GAO to examine FDA's (1) monitoring (sampling and testing) of the nation's domestic food supply for illegal pesticide residues and (2) efforts to prevent food containing illegal pesticide residues from reaching the market.

Background:

The Federal Food, Drug, and Cosmetic Act gives FDA responsibility for prohibiting the interstate marketing of food containing illegal pesticide residues. Through its pesticide monitoring program, FDA collects samples of domestic food and tests each sample for certain pesticides primarily by using tests (known as multiresidue methods) capable of detecting large numbers of pesticides on a food sample. When FDA finds illegal residues in the food, the act gives FDA authority to prohibit it from being marketed through seizures or injunctions, and to seek criminal penalties against those who market adulterated foods.

Results in Brief:

FDA has concluded that it cannot monitor all food that might contain illegal pesticide residues. Consequently, FDA has designed its monitoring program to act as a deterrent by selectively spot-checking a very small amount (probably less than one percent) of domestically produced food for illegal pesticide residues and to remove such food that it finds to contain such residues.

FDA's pesticide monitoring program as it is currently carried out has two major shortcomings:

• FDA does not regularly test food for a large number of pesticides that can be used or may be present in food. Included among these are a number of pesticides that, according to FDA, require continuous or periodic monitoring because they are known as potential health hazards and are likely to be used.

• FDA does not (1) prevent the marketing of most of the food that it finds to contain illegal pesticide residues and (2) penalize growers who market food with illegal pesticide residues when FDA is unable to remove it from the market.

GAO is proposing a matter for congressional consideration and making a recommendation that should help to provide a more effective program to protect the public from adulterated food.

Penalties Are Not Being Assessed for Marketing Adulterated Foods:

For FDA to effectively enforce pesticide residue limits, it must be able to remove domestic food containing illegal pesticide residues and penalize growers who ship such food and do not remove it from the marketplace. In 107 of 179 pesticide violations GAO reviewed, FDA did not remove any of the food and in no cases did FDA penalize growers who shipped the food. FDA's position is that it does not remove more food or penalize growers because existing legislative authority is not well suited for pursuing pesticide violations on domestic food. This is because FDA (1) cannot detain domestic food containing illegal pesticide residues while it seeks court action to remove it from the market and (2) must rely on criminal penalties that require extensive evidence gathering and are costly to prosecute because FDA does not have the authority to impose civil penalties on growers who market adulterated food. GAO has previously proposed that FDA be granted detention authority and continues to support this proposal.

Want The Pesticide Industry in Your Milk?

BY ALBERT H. MEYERHOFF

THE NEW YORK TIMES, MAY 26, 1987

As hearings begin this week on proposed reforms of the Federal pesticide law, members of the House Agriculture Committee should finally stand up to the powerful business lobbies that for nine years have prevented passage of a statute that would protect human health and the environment.

The pesticide industry deservedly suffered sharp criticism recently for its record of failing to consider public health. Pesticides, which are intended to terminate insect life, are among the most deadly substances known. Many have been linked to cancer, birth defects, genetic mutations and other serious disease. A seemingly endless chain of debacles involving these chemicals has generated renewed public concern.

Last week, the National Academy of Sciences released a study that concludes that 28 pesticides that are known to cause cancer and that have been found in foods might cause more than one million cancers over the next 70 years.

In 1983, much of Hawaii's milk supply was found to be contaminated by heptachlor, a chemical that was banned in the rest of the country in the early 1970's because it was linked to leukemia and neurological disorders.

In 1984, the nation's grain supply, citrus crops and other foods were found to contain the powerful chemical ethylene dibromide, which causes cancer and birth defects. Last year, consumers learned that some supplies of apple juice may have been tainted with a "growth regulator," daminozide, which was previously linked to cancer in tests conducted by the National Cancer Institute.

Moreover, evidence of pesticide contamination of drinking wells is growing. In California alone, more than 2,000 wells in 23 different counties were discovered to contain 57 different pesticides, including nearly 1,500 wells now closed because of contamination from a known human carcinogen. Nationally, the first serious monitoring of drinking water for pesticide contamination is now going forward. Preliminary results indicate a major problem in at least 20 states.

During the last session of Congress, the House and Senate agriculture committees had by far their best opportunity in more than a decade to accomplish obviously needed reforms. Representatives of the pesticide industry had grudgingly agreed to negotiate with a coalition of environmental, labor and consumer organizations on a new Federal pesticide law.

To the surprise of many, agreement was finally reached on a sweeping pesticide reform bill, introduced by the chairmen of the Senate and House agriculture subcommittees.

Separate, successful negotiations took place with the American Farm Bureau, an industry organization, on agricultural issues and the proposed compromise would have made several advancements, including these:

• Requiring full health and safety testing of all pesticides in accordance with a rigid timetable.

Fewer than 10 percent of the 2.5 billion pounds of pesticides sold annually have been adequately tested for health effects, according to the National Academy of Sciences.

• Establishing strict controls to protect the nation's drinking water from contamination.

• Prohibiting imports of food containing residues of banned pesticides.

• Streamlining cumbersome pesticide regulatory procedures.

• Imposing an industry user fee that would generate some $75 million for the Environmental Protection Agency to better protect the public.

Unfortunately, the agriculture committees allowed

the bill to be ensnared by a variety of unrelated amendments at the behest of other special interests, including the food industry. Finally, time simply ran out in the 99th Congress.

With the new 100th Congress and a Senate under Democratic control, the chances to enact comprehensive reform legislation have improved significantly.

Building on last year's efforts, and armed with new and alarming data from the National Academy of Sciences, this Congress has the opportunity to pass even stronger legislation. But if such reform is to occur, the efforts of the food industry and farm organizations to use the Federal pesticide law to pre-empt states that are considering their own laws or to escape liability under other Federal laws must be decisively turned aside. Only then can the public finally be protected from these highly dangerous chemicals. ◆

Albert H. Meyerhoff is a senior attorney with the Natural Resources Defense Council.

Getting Off the Pesticide Treadmill

BY MICHAEL DOVER

TECHNOLOGY REVIEW, NOV./DEC. 1985, PP. 53+.

Humanity has always had to contend with microbes, plants, and animals that threaten our health and comfort, damage our property, harm our domesticated animals, and encroach upon our food supply. Only in the last four decades has the relationship between people and pests shifted dramatically, raising our expectation that the battle could be won. This revolution – for it is nothing less than that – began during World War II with the development of DDT, the first of the synthetic organic pesticides.

With its low toxicity to humans and high toxicity to insects, DDT seemed to be a miracle chemical – the perfect pesticide. It clearly was safer and more consistently effective than the highly toxic arsenicals, heavy metals, nicotine, and cyanide used to control insects earlier. DDT had other apparent assets as well, including persistence in the environment and toxicity to a broad spectrum of insect species. And the chemical worked: it prevented millions of deaths from lice-borne typhus during and immediately following the war, and it virtually eliminated malaria, carried by mosquitoes, from large parts of the world. Together with the later discoveries of a burgeoning chemical industry, DDT also revolutionized agriculture. Because farmers no longer had to worry as much about pest control, they could concentrate on reducing labor and using fertilizers, irrigation, machinery, and high-yielding crop types. These changes have profoundly affected the world's economies and societies.

Yet while DDT initiated the new age of pest control, it also spawned a new environmental consciousness. For DDT became the principal villain in the problems that emerged as our society began to rely almost exclusively on chemicals for pest control. Soon after the chemicals were developed, questions about their effects on human health and the environment, and about pests' resistance to their effects, began to surface.

These concerns have proven to be well founded. Many species of insects no longer respond to the effects of chemical pesticides. World pesticide use has increased dramatically – U.S. production rose from 464,000 pounds in 1951 to 1.5 billion pounds in 1980 – but the percentage of crops lost to pests has apparently not declined. Insects consume as much as one-third of the Asian rice crop annually, and the United States losses of fruit and vegetable crops from plant diseases may exceed 20 percent. Malaria is again becoming a serious health concern in the tropics. And reports of pesticide contamination of water and food supplies surface regularly. Clearly, just pouring on more chemicals is no answer.

If agricultural productivity is to meet rising food demand and insect-borne diseases are to be controlled in tropical areas, safe, effective pest control must become a global concern. And indeed, better pest-management techniques stand ready to reduce pesticide use by as much as 50 to 75 percent. Yet our society continues to rely almost solely on chemicals. We are caught in a "magic-bullet" approach, continually trying to develop new pesticides to stay one step ahead of disaster.

I believe we can better manage pests by carefully deciding when, where, and how much pesticide to use. We must also employ many different technologies, including biological controls, as part of a long-term strategy. The key is to base the choice of technology on an understanding of how pests interact with one another and the environment. That will be possible only if we abandon our unsystematic efforts and set up a comprehensive federal and state research and advisory program to help farmers and others manage pests safely and efficiently.

Problems with Pesticides

By design, pesticides are biologically active and, in most cases, toxic. Thus, they pose potential risks to human beings. However, 40 years of experience have provided remarkably few consistent results on how these materials affect people. Of most immediate concern is acute poisoning – sickness and death from one or a few exposures to a chemical. Although scientists debate the incidence of such poisonings, estimates of 10,000 deaths and 400,000 illnesses per year worldwide are probably close to the mark. Continuing exposure to low levels of pesticides also puts farm workers at risk of tumors, reproductive disorders, birth defects, and long-term illnesses, including cancer.

Those who handle pesticides are not the only group at risk. Pesticide residues in food and drinking water may accumulate in human tissues, causing cancer and other diseases. Especially disturbing are the residues' effects on infants and children, a subject scientists and regulators have neglected. Because so many people are potentially (and involuntarily) at risk, dietary exposure remains a powerful social and political issue. The result has been public pressure to ban certain pesticides. In the 1970s the EPA banned many of the "organochlorine" insecticides – DDT and its relatives – from general use because of public outcry over their residues in humans and wildlife. Residues of these chemicals in food have since dropped significantly.

Yet the ecological and health effects of pesticides continue to plague us. The chemicals that have replaced the organochlorines – such as the organophosphates, which include malathion and parathion, and the carbamates, which include carbaryl and aldicarb – are less persistent. But they are also more water-soluble and therefore more likely to reach aquatic and marine organisms. Many of these materials are more acutely toxic as well, posing a greater hazard to farm workers in particular. Moreover, some do not break down as quickly as scientists had originally thought, making long-term effects on "nontarget populations" a serious possibility.

Heavy reliance on synthetic pesticides has also spawned problems that jeopardize effective pest control itself. For example, many species of pests are becoming immune to the effects of these chemicals. From 1970 to 1980 the number of such resistant species of arthropods – insects, mites, and ticks – jumped from 224 to 428. The numbers of resistant species of rodents, bacteria, fungi, and weeds are also increasing. Resistance threatens the future of potato farming in the Northeastern United States and cotton growing in several parts of the world. In the tropics, resistance also impedes control of insect-borne diseases: the chemical industry can no longer develop new pesticides as quickly as pests develop immunity.

Another problem stems from the fact that both on farms and in natural habitats, predators, diseases, and competitors keep pests in check. Pesticides may disrupt this balance so that once-harmless species grow numerous enough to become pests. The burgeoning of the tobacco budworm on cotton crops in Texas and Mexico, and the explosion of spider mite populations on apples in the Pacific Northwest, are among the best-known examples. At worst, these secondary, "induced" pests cause more trouble than the pests that the chemicals were originally designed to control. Furthermore, because natural checks have been eliminated, primary pest populations may rebound quickly to even greater numbers after a pesticide wears off.

Resistance, secondary pest outbreaks, and resurgence often lead farmers to apply more and more chemicals to keep from losing everything. Just such a "pesticide treadmill" devastated cotton growing in parts of Central America in the 1950s and 1960s. In little more than 10 years, the number of insect species requiring control went from two to eight, while the number of pesticide applications per growing season increased from fewer than five to twenty-eight. Pest-control costs eventually accounted for half of all production expenses.

Pesticides present an ever-widening dilemma. While they have opened up many possibilities for improving agriculture and public health, they have closed others, making us extremely dependent on chemicals for continued survival. Biological controls, which are more benign and often more effective, are crucial to any strategy that would relieve us of this dependence.

Putting Nature to Work

In agriculture and other managed ecosystems, people have long used "natural control," a form of biological control that involves conserving pests' natural enemies. Farmers keep these beneficial species active by carefully selecting their pesticides and by judiciously timing planting, harvesting, and pesticide use. For example, apple growers control spider mites by protecting other mites that prey on the pests. To do this, growers avoid insecticides and fungicides that are toxic to the beneficial mites, and often maintain ground cover at the base of trees to protect the beneficials during winter.

Paradoxically, the future of natural pest control depends partly on developments in pesticide chemistry. For instance, new insecticides might target key pests more specifically, allowing predators and parasites to survive to suppress secondary pests. Or newly designed chemicals might have a composition that allows only beneficial species to detoxify them. However, innovations may be possible only if scientists find ways – using genetic engineering, for example – to synthesize complex molecules cheaply.

A second form of biological control, "inoculative release," entails introducing pests' natural enemies and making them a permanent part of the ecosystem. One of the earliest examples of such an effort occurred in 1888, when California imported the vedalia lady beetle to feed on the cottony-cushion scale, a pest that threatened to wipe out the state's citrus crop. Since then, the pest has generally remained innocuous, making occasional comebacks only when insecticides have been overused. This spectacular success triggered biological-control programs in many parts of the world, and so far at least 327 introduced natural enemies have successfully controlled pests.

The technique's payoffs in crops saved and chemical costs foregone are virtually limitless. For example, introducing Chrysolina beetles to control a California range weed toxic to cattle cost about $750,000 in the 1940s, and savings have reached more than $100 million. Also, because biologists choose their target pests and beneficial species very carefully, the success rate for inoculative release is remarkably high – 55 to 75 percent of the introduction programs have been effective. And because of the extreme care scientists have used in screening natural enemies before release, the environmental cost of this approach has been almost zero.

Biological agents need not become permanent parts of the ecosystem to be effective. "Inundative release" entails periodically blanketing an area with beneficial species of insects, microbes, and other natural enemies – living pesticides – to bring pest numbers down to tolerable levels. For instance, U.S. soybean farmers effectively and economically control the Mexican bean beetle by annually releasing a parasitic wasp that the U.S. Department of Agriculture (USDA) supplies at cost. These biological agents later succumb to natural causes, such as seasonal climate change.

Like chemical pest control, inundative release involves continual expenses for production, storage, transportation, and application. This means that unlike inoculative release, which requires public funding to support a one-time, regional effort, inundative release is a likely target for commercialization. In many parts of the world, private firms sell pest-controlling microbes – most notably Bacillus thuringiensis (BT). BT enjoys widespread popularity as a means of controlling various types of caterpillars, and a new variety of BT has proven very effective against mosquitoes and blackflies. These inundative biocontrols are sometimes more expensive than chemicals, but the approach is often cost-effective for high-value crops such as tree fruit and greenhouse crops. The approach also becomes attractive when the risks of human exposure to chemicals are unacceptable.

New techniques in genetic engineering and tissue culture may make wider use of this type of biocontrol possible. Monsanto Corp. has developed a genetically engineered bacterium that controls rootworm when applied to corn seeds – tapping what is probably the biggest U.S. insecticide market. The bacterium will be available for commercial use only if it passes extensive safety tests required by the Environmental Protection Agency (EPA). Other biotechnology companies are developing new strains of BT that would work against more insect species. Whether these altered microbes will pose health and enviornmental risks is a matter for serious concern, and the EPA is beginning to develop criteria for registering and testing such biocontrols.

The advantages of using biological controls are great: the benefit-cost ratio for inoculative release can be well over 100 to 1, according to a report prepared for the World Bank. By comparison, an independent analysis at Cornell University puts the average benefit-cost ratio for chemical control at 3 or 4 to 1, depending on environmental costs. Yet the full potential of biological controls to reduce chemical costs, increase yields, and enhance environmental quality is still unrealized. This unfulfilled potential seems particularly unfortunate in the poorest countries, many of which spend precious foreign exchange to gain the dubious short-term gains of chemical-based control (often at the expense of the public's health).

A leader in the field of biological pest control, Paul DeBach of the University of California, has written, "As long as we ignore, anywhere in the world, an effective natural enemy capable of controlling one of our major pests, we are postponing cheap, reliable, and permanent control." We may also be sacrificing public health and environmental quality every time we opt for a potentially hazardous chemical over a benign biological agent. But further research is necessary: how much we can take advantage of pests' natural enemies ultimately depends on how fully we understand their ecology and behavior.

Cheap and Effective: Host Resistance

Over their long evolutionary history, plants have developed host resistance: ways to ward off almost every insect and pathogen that has attacked them, as well as means to compete with other plants for water and nutrients. Whether through trial and error or with the aid of modern breeding techniques, farmers have always selected crop varieties with superior host resistance. Growing resistant crops is often the easiest, cheapest, and most effective way to control insects and diseases. Indeed, for some pests, especially plant viruses, there are no known alternatives.

As much as 75 percent of U.S. cropland is planted with disease-resistant strains, creating a net benefit exceeding a billion dollars. In Asia, resistance bred into rice varieties helps account for farmers' dramatic productivity gains over the last two decades.

Resistance allows fewer troublesome insects and pathogens to survive. It also slows down pest development so that populations do not peak during the growing season and farmers can delay pesticide use, conserving natural enemies. In addition, breeding can create crops whose planting, maturation, and harvest times do not coincide with pests' natural cycles. For example, growers' use of short-season cotton has done as much as any other control method to keep insects in check.

According to the U.S. Council for Environmental Quality, the benefit-cost ratio for efforts to develop host resistance is about 300 to 1, not including reduced pesticide costs. But research on host resistance must take a new form, since farmers are devoting more and more land to major crops. These crops are often of one genetic type, and global agriculture is therefore becoming more vulnerable to new strains of pests. For instance, in 1970 an epidemic of leaf blight struck the Corn Belt because large areas were planted with the same genetic type. Once the pathogen evolved a way to overcome the type's resistance, the entire "genetic lawn" lay open to infection. Losses were crippling.

Farmers can avoid such disasters by planting mixtures of similar genetic types that differ mainly in their resistance characteristics. Such "multiline" mixtures have halved the disease rate from rust fungi in oats. If host resistance is to remain a viable strategy, plant breeders and pest-management specialists must begin to consider not only how resistant plant types are but also how quickly pests can change to overcome plants' defenses. Too often, the tendency has been to try to stay just one step ahead of pests in this evolutionary race.

However, research on plant breeding may well be narrowing rather than broadening in scope. Progress in developing resistance has usually occurred in university and government laboratories, where scientists have been eager to publish their results so that all could benefit. Yet the USDA is funding less academic and government research, and breeding is becoming more commercialized. Competition will undoubtedly make plant-breeding companies reluctant to publish their results. Because of patenting, genes incorporated into one plant variety may be unavailable to breeders trying to improve resistance in another variety. Efforts to develop multilines and special resistant plants may lose out to work on plant characteristics that many farmers demand, such as improved response to fertilizers and greater ease in mechanical harvesting. Short-term goals may take precedence over long-term progress.

The only way to avoid these dead ends is for the USDA and universities to maintain a vigorous public plant-breeding program. It is especially important that public funds meet the needs of growers of fruits, vegetables, and other smaller-market commodities that may not attract enough private research capital. Commercial breeders may bring what is already in demand to the market quickly. But without public support for exploring what is not yet known, the private sector's apparent efficiency could be short-lived and our options for protecting crops and animals severely limited.

Putting the Pieces Together

The full potential of biological controls and plant resistance can be tapped only through a carefully coordinated pest-management program. Conversely, without such a program, biological controls could well fail. Improper or untimely use of pesticides could destroy biological agents. A resistant crop variety could decimate pest populations, starving predators and parasites important to natural control. Microbial pesticides may not work well unless carefully timed for maximum effect. In short, each technique's special characteristics deserve careful examination because of their potential effects on other elements of the pest-control program.

In the late 1960s and 1970s, mounting evidence of the environmental hazards, health risks, and costs associated with indiscriminate pesticide use sparked interest in integrated pest management (IPM), just such a systemic, long-term strategy. Unlike conventional approaches to pest control, IPM is based on a sound understanding of the crop, its pests and their natural enemies, competitors, and alternative hosts. IPM emphasizes "systems design" rather than knee-jerk reactions, aiming to stabilize pest-control systems and improve the predictability of control. Instead of trying to eradicate pests, IPM practitioners use a variety of tactics to maintain them at harmless levels. The IPM approach also seeks to balance the numerous costs, risks, and benefits involved in managing pests. Those who adopt IPM in their own best economic interests bring environmental benefits to all.

One of the first U.S. IPM programs began in the 1960s at Texas A&M University to fight the boll weevil and the pink bollworm, key pests of the south-central Texas cotton crop. Excessive spraying of insecticide had so disrupted natural controls as to make serious pests of another species of bollworm and the tobacco budworm – both previously innocuous. Although more chemicals temporarily kept these new pests at bay, the tobacco budworm eventually developed such high resistance that farmers faced ruin.

The IPM program has provided a more effective system for managing pests. The main lines of defense include the use of short-season cotton, which helps control the boll weevil, and the elimination of second and third cuttings, which ends the growing season before the pink bollworm population can build up. Farmers also coordinate irrigation and planting times and plow under crop remnants to deny insects food and shelter at critical times. "Scouts" – working either for state extension services or as private consultants – monitor the population levels of pests and their natural enemies so that growers apply insecticides only when absolutely necessary.

These efforts have yielded dramatic results. In two counties where harvested cotton acreage had dropped from about 105,000 in 1970 to about 50,000 five years later, harvested acares increased to 236,500 by 1980. Farmers netted benefits estimated at $29 million from higher yields and lower production costs.

Illinois has also developed an IPM program to control corn rootworm. The EPA banned the principal insecticides heptachlor and chlordane in the 1970s, when unacceptable environmental and health risks cropped up. But the rootworm had been growing resistant to these chemicals for years anyway. Today farmers monitor pests, rotate crops, and selectively apply other insecticides. In fields where August rootworm counts are high, farmers plant soybeans or other rootworm-resistant crops the next year. Where fields are left in corn, farmers apply a soil insecticide only if each plant averages more than one rootworm. As a result, pesticide use has decreased, and insects are not becoming resistant. Also, the soybeans are conserving soil nitrogen, and the crop rotation is helping keep some weeds at bay. Even farmers who do not rotate their crops derive some benefit from the area's overall decline in rootworms.

Chemical pesticides are still the main means of control for apple growers, since pests attack the fruit itself and farmers hoping to sell in the fresh-fruit market can tolerate almost no damage. Yet even here long-term management strategies have reduced dependence on routine spraying. For example, Michigan has pioneered the use of an extensive system of agricultural weather stations, computer terminals, scouting services, mathematical models, and telecommunications to advise apple growers of prevailing pest conditions. These techniques help determine when insect eggs hatch and when adult males emerge from dormancy, so that farmers can carefully time their doses of insecticide.

The principles of IPM work equally well for managing insects, plant diseases, and weeds in forests, rangeland, pastures, suburban yards, and urban parks. The tenets are also effective against pests that attack livestock and people. The key is knowing as much about a pest and its environment as possible, and using this knowledge to determine when to intervene in the pest's life cycle.

Building an IPM Support System

IPM got its initial push in the United States from the federal government. The EPA, USDA, and National Science Foundation jointly funded a 17-university pest-management research effort in the early 1970s. This program, the Huffaker Project, laid the groundwork for using "systems science" – the basis of IPM – to solving pest problems. The Huffaker Project's successor, the 16-university Consortium for Integrated Pest Management, has made impressive gains in developing IPM programs for alfalfa, apples, cotton, and soybeans. Spurred in part by these results, the USDA began an extension program, now national in scope, to show farmers the benefits of IPM. The result is that a new industry of private IPM consultants has sprung up: scientist-entrepreneurs who provide monitoring and advisory services to growers of cotton, corn, fruits, and many other crops. Meanwhile, the USDA has been putting together an interdisciplinary research program to develop regional projects for IPM.

However, IPM has fallen on hard times since 1981. Along with many other programs, IPM has suffered from Reagan budget cuts. The EPA, the initial funder of ground-breaking IPM research, now has no IPM research program at all. The USDA's extension program has had to fight off yearly attempts to close it down, and the regional research projects in IPM have barely begun despite years of planning. A promising initiative of the Carter administration – a plan to make federal agencies adopt IPM whenever they use pesticides – died almost as soon as the White House changed hands. Meanwhile, at the state level, many research and extension administrators are acting as if IPM were a thing of the past.

Some of these setbacks are more than political in origin. Since the problems of an entire ecosystem are difficult to tackle and easy to avoid, IPM has lost out to a rebirth of the magic bullet philosophy. An appealing new generation of pesticides – the synthetic pyrenthroids – has come along in the last decade. Many ecological concerns voiced in the 1960s and 1970s have faded in the rush to join the coming biotechnology revolution, which offers the prospect of yet another round of "breakthroughs" in pesticide development and plant breeding. These changes have helped spur today's private-product approach in agricultural research policy.

The attraction of relying on individual technologies, especially those developed by private firms, is obvious. Manufacturers provide an army of advisors – salespeople and technical-support staff – to farmers at no apparent cost to the taxpayer. In contrast, IPM requires an extensive, often publicly supported, infrastructure. Moreover, there are no economies of scale in providing IPM advisory services and field-level monitoring, unlike in manufacturing. Extension agents who teach sound ecological principles to the next generation of farmers and pest-management consultants can promise only long-term rewards. Proponents of the biotechnology boom, on the other hand, have been promising lucrative near-term pay-offs.

But leaving pest management to market forces invites a new round of "surprises" from pest resistance and resurgence, and from outbreaks of secondary pests. Rather than waiting to be surprised, policymakers should begin programs to ensure that the most-needed technologies are developed and used.

When a company or university develops a new technology, federal and state agencies should carefully weigh its potential for disrupting present pest management. For instance, the EPA now pays little attention to the possible effects of pesticides on beneficial species. The agency should require chemical companies to test for these effects before it approves pesticides for general use.

Where market forces alone cannot foster sound pest-management technologies, government action is essential. For instance, the USDA should fund more programs to introduce beneficial species. Methods of predicting how insects become resistant to pesticides are crude at best, as are means for anticipating the durability of host resistance. Again, federally supported research is neccessary. In addition to funding programs to develop new varieties of pest-resistant crops, the USDA should set standards for pest resistance that all breeders must meet. The USDA could also provide more support for research on using biotechnology to improve pest control.

How should the government finance this new IPM infrastructure? A public fund based on a nominal sales tax on pesticides – two cents per pound, for instance – would raise about $20 million a year, enough to defray most of the costs. By comparison, the entire EPA research budget for IPM from 1970 to 1980 was only $19 million. Commodity associations and farmer cooperatives could cover some of the rest, funding crop-specific research and monitoring activities. Of course, in today's political climate, Congress is unlikely to consider any new tax, however modest, to fund government programs. But a pesticide tax for IPM is just the sort of "pay-as-you-go" system – in which the beneficiaries are the principle source of funds – that makes sense.

Just as the airline industry depends on publicly supported weather monitoring and forecasting, air-traffic control, and safety research and regulation, so too do pest control and its associated industries require the research, monitoring, and advisory services that only the public sector can provide. And just as human health-care systems require effective national management, so do plant-health systems. Without such support and management, progress in developing sustainable, ecologically sound pest control will be sporadic at best.

Technology Review is published 8 times a year. Subscriptions are $24.
Technology Review
Building W 59, MIT Cambridge MA 02139

Michael Dover is an ecologist & author of *A Better Mousetrap: Improved Pest Management for Agriculture*, World Resources Institute, publisher.

Editorial:
Foodicides
WASHINGTON POST, MAY 26, 1987

In 1972 Congress passed legislation to modernize the way the government regulates pesticides. The job was given to the new Environmental Protection Agency. Astonishingly little has happened since.

Most of the pesticides in use today were already on the market when the new law was passed. Many would fail contemporary tests, and EPA was supposed to review them. It never has. Bills to quicken the process have failed. Last year environmentalists finally seemed to have struck a deal with chemical manufacturers on a relatively simple bill to set a schedule. That too failed as disputes cropped up over side issues: pesticide patent terms, liability questions and the rights of states to set regulatory standards of their own.

This year the issue is back, with a difference: there is talk not just of setting a schedule but also of adjusting the standards by which the pesticides are judged. It is in this context that a new report by the National Academy of Sciences, commissioned by EPA, was issued last week. The report reminds us that the law, an accretion of many decisions over the years, is neither very tidy nor entirely rational. It sets one standard of safety for raw foods, a higher one for processed. It has a strange definition of processed, so that certain foods such as milk and eggs, even though they are "processed," are exempt. The unevenness has been compounded by the difference in treatment over the last 15 years of new pesticides and old. Because new pesticides have been subject to closer scrutiny, some of them have been barred that are safer than some older ones that have been ignored.

The result is that we take some poison with our food (and occasionally with our drinking water as well). How much finally reaches the dinner table and exactly how dangerous it is are not clear, but it's plainly a serious problem. The report suggests that in finally addressing this, EPA might benefit from a uniform and slightly more flexible standard for all exposures. Maybe so, but our sense is that the problem is not the elegance of the standard, but its application. If the law is clumsy, the authority nevertheless exists to provide much more protection than EPA, under several administrations, has provided. It's fine to seek a better club, but meanwhile EPA should swing the one it has. ◆

Editorial:
Government Must Make Food Safer

USA TODAY, APRIL, 29, 1987

"It must be something I ate."

Next time you hear that phrase, or moan it yourself, take a moment to consider what might have caused that upset stomach. It just might be the germs, pesticides, or drugs that came to dinner with your food – uninvited:

• It might be antibiotic-resistant germs.

More than 20 million of us come down with food poisoning each year. Most of us get over it, but 9,000 die annually. And an increasing number of food poisoning cases – more than a quarter million each year – are caused by germs, resistant to antibiotics, consumed in meat from cows, pigs or chickens raised on antibiotic laden feed.

• It might be forbidden pesticides.

Some pesticides hurt more than weeds and insects – they can cause birth defects, nervous disorders and cancer in humans. Our government has banned the worst of them, but many foreign governments have not. Result: Imported fruits and vegetables laced with forbidden chemicals.

Mexico alone provides half the produce we eat each winter. Yet government inspectors check only 1 percent of imported food shipments. And usually, by the time illegal pesticides are found – as happens 6 percent of the time – the rest of the shipment has already been marketed and eaten.

Domestic fruits and vegetables sometimes are tainted, too, with excess levels of legal pesticides. But the government checks only a fraction of domestic shipments.

• It might be body-altering drugs.

Hormones, which regulate reproductive activity in humans, are commonly fed to livestock to stimulate growth. The European Economic Community has voted to ban this practice, blaming hormone-laden meats for an outbreak in Italy of sexual abnormalities in children. Experts say hormones will do no harm if properly used in livestock. But the National Academy of Sciences says our government system for detecting excess hormones in meat is inadequate.

We needn't be afraid of everything on our plates, or even most of it. But we should be cautious.

We can urge our supermarkets to carry contaminant-free meats and produce. Some grocery chains already do.

We can change our cooking habits. Scrub produce thoroughly. Cook meats well to kill germs. And clean cutting boards between uses so that salmonella bacteria from chicken, for example, won't spread to salad greens.

But some poisons can't be scrubbed or cooked away. We need the government to protect us from those. Congress is listening: Hearings start tomorrow on pesticides in food.

But we need more: The Food and Drug Administration must find and stop tainted shipments. The Environmental Protection Agency must set better standards for pesticide residues. And Congress must pass a tough insecticide law to replace the one that is expiring. Fines against violators and fees for pesticide manufacturers can foot the bill.

Our kitchens aren't chemistry labs. When our families sit down to dinner, they should find wholesome food on their plates, not antibiotics, pesticides and poisons. ◆

Everything Doesn't Cause Cancer
But How Can We Tell Which Chemicals Cause Cancer and Which Ones Don't?
U.S. DEPARTMENT OF HEALTH, EDUCATION, AND WELFARE PUBLIC HEALTH SERVICE

NATIONAL INSTITUTES OF HEALTH

Question: What causes cancer?

Answer: Cancer is actually many different diseases with many causes. Most human cancers probably are caused in part by environmental factors. Cancer-causing agents, called carcinogens, include certain man-made and natural chemicals which may be found in small quantities in air, water, food and the work place. Cancer-causing agents also include x-rays, sunlight, and certain viruses. Contact with cancer-causing agents may result from individual actions, such as smoking or dietary habits. This brochure deals mainly with chemical carcinogens. It is not true that everything causes cancer, or that the problem is hopeless. Relatively few substances cause cancer. Most chemicals, even most toxic or other dangerous chemicals are not carcinogenic. Susceptibility to carcinogens may vary among individuals.

Question: Can cancer be prevented?

Answer: Yes, many cancers could be prevented by reducing our exposure to carcinogens. Some carcinogens can be avoided by personal choice, government regulation, corporate decisions or other societal actions. Reducing human cancer rates by reducing or avoiding exposure to cancer-causing agents is an achievable goal.

Question: How soon after exposure to a carcinogen does the cancer appear?

Answer: Cancers develop slowly in man, usually appearing 5 to 40 years after exposure to a cancer-causing agent. Cancers of the liver, lung or bladder, for example, may not appear until 30 years after exposure to vinyl chloride, asbestos, or benzidine. This long latent period is one reason why it is so difficult to identify the causes of human cancer.

Question: How can we identify agents that cause cancer in people?

Answer: It is hard to do so directly because suspected carcinogens are tested on laboratory animals, not people. But direct human exposure to cancer-causing substances has often occurred, nevertheless, and we can study the exposed populations. For example, people who have been exposed to tobacco smoke or asbestos develop after many years a higher frequency of cancer of the lung and other organs than unexposed people.

From such population studies we have identified about 30 agents as human carcinogens.

Questions: Aren't there other ways to identify carcinogens, without humans first getting cancer?

Answer: Yes, tests on laboratory animals can identify substances that are likely to be human carcinogens. Mice or rats are most commonly used for such tests because they are small, easily handled, more economical than larger animals, and similar to humans in their response to carcinogens, at least in a general way. Most major forms of human cancer have been reproduced in such animals through exposure to chemical carcinogens. Since their natural lifetime is two to three years, rodents provide information about the cancer-causing potential of test materials more quickly than do longer-lived animals, such as dogs or monkeys. Special strains of mice and rats have been developed to be particularly suitable for carcinogenicity testing.

Question: How well do laboratory animal tests predict whether or not a substance can cause cancer in humans?

Answer: There are two ways to answer that question. Of the approximately 30 agents known to cause cancer in humans, almost all cause it in laboratory animals.

Of the several hundred other chemicals that cause cancer in laboratory animals, however, it is not known how many are also human carcinogens. Nevertheless, materials that cause cancer in one mammalian species are usually found to cause cancer in other species. Furthermore, in some instances the risk predicted in advance by tests on rodents were later confirmed by the occurrence of cancer in exposed humans. Chemicals such as diethylstilbestrol (DES), vinyl chloride, and bis(chloromethyl)ether were shown to cause cancer in mice and rats before it was known that people exposed to those chemicals also had increased cancer rates.

We should assume, therefore, that agents that cause cancer in animals are likely to cause cancer in humans. To prevent cancer, we cannot afford to wait for absolute proof of carcinogenicity in humans. Instead, we must heed the warnings provided by laboratory animal experiments and reduce or eliminate human exposure to probable cancer-causing agents.

Question: How are these laboratory animal tests performed?

Answer: In brief, groups of about 50 mice or rats of each sex are exposed to the test substance at different dosages for about two years. Other groups, known as controls, are treated identically, but are not exposed to the test substance. At the end of the experiment, the animals are carefully dissected and examined by pathologists (doctors who interpret the changes in body tissue caused by diseases), and the frequency of tumors in the test groups is compared with that of the controls. Carcinogens produce a tumor frequency higher in the exposed animals than in the unexposed control animals. Non-carcinogens, by contrast, do not produce tumors.

Question: We often read abut mice or rats being fed dosages much higher than those to which humans would normally be exposed. Are high doses really used, and if so, why?

Answer: Yes, high doses are often used to increase the ability of the tests to detect tumor-causing potential. The public often misunderstands the reasons for high dosage testing, and misinterprets the results.

In the human population, large numbers of people are exposed to low doses of chemicals, but the total impact may not be small at all. For example, a carcinogen might cause one tumor in every 10,000 people exposed to it, which may seem like a needle in a haystack. But exposure of 220 million Americans would result in 22,000 cancers - a public health disaster. We therefore need sensitive tests to detect those agents with the potential to cause only low cancer rates.

We obviously could not identify a carcinogen that causes one cancer in every 10,000 exposed mice by running the test on only 50 mice. To detect such a low cancer rate, we would need tens of thousands of mice, which would cost many millions of dollars per test. Testing more than a few chemicals in such an unwieldy fashion would be prohibitively expensive and time-consuming.

With high dosages any potential carcinogenic effects are more likely to be detected in small groups of rodents because the cancer rate among the test animals is increased correspondingly. In the above example, a dose 5,000 times higher might cause cancer in 5,000 of every 10,000 mice, or 50 percent of the animals. If 20 or 30 of our test group of 50 mice developed cancers under such conditions, while the control group had only a few cancers, we could properly conclude that the chemical was capable of causing cancer. When high doses do not cause cancer, we also have greater assurance that the chemical is not a carcinogen.

Question: Won't the animals get sick and die if give such high doses?

Answer: No. National Cancer Institute guidelines for testing restrict the doses to levels that will not cause significant toxicity or unduly shorten the animal's lives. Because the animals must live long enough for tumors to develop, doses that kill the animal prematurely are not used in tests for carcinogenicity.

Question: Doesn't everything cause cancer if the dose is high enough?

Answer: No. High doses of many chemicals are toxic, but they will not cause tumors. Other forms of toxicity, such as loss of hair or weight, various organ malfunctions, or even death, should not be confused with carcinogenesis.

In one study, 120 pesticides and industrial chemicals were tested at the highest doses the mice could tolerate and survive. These chemicals were not randomly selected, but were chosen because they were suspected of carcinogenicity. Nevertheless, after two years of such treatment, only 11 of these chemicals caused cancer in the test animals.

Question: Will a carcinogen cause cancer in the same organ when tested in different animal species?

Answer: Often, but not always. Different organs may become cancerous in different species. Thus a chemical that causes cancer of the liver in mice, for example, might cause cancer of the breast in rats and cancer of the bladder in humans. A carcinogen also can cause tumors in several organs in the same species, or even in the same individual.

Question: Why are test animals sometimes exposed to the test chemical in a way different from the way in which humans would be exposed? Hair dyes were fed to rats, for example.

Answer: The method of exposure need not be identical, although that is usually preferred. The tests seek to determine the potential to cause tumors, and distribution throughout the animal's body is the important part of the test. Hair dyes can enter the circulatory system through either the scalp or digestive tract.

Question: Aren't many of the animal tumors not cancerous?

Answer: Yes, but these tumors also indicate human cancer risk. Noncancerous (benign) tumors often become cancerous. No chemical is known to cause only benign tumors.

Question: How should we interpret the appearance of tumors in one species, such as rats, and the absence of tumors in another species, such as monkeys?

Answer: Positive evidence of tumors in one test is not cancelled by negative evidence, or the absence of tumors, in another test. A substance that is carcinogenic in one strain or species occasionally has little or no effect in another strain or species. Furthermore, not all tests are equally sensitive, and a negative result may mean merely that the effect was missed. Too few animals, too low a dosage, or too short a test period, for example, can lead to a false negative result. A 5-year test on 10 monkeys could fail to detect tumors that would have appeared after 10 to 15 years, or that would have been detectable in 50 monkeys. Any adequately performed positive test in laboratory animals indicates a cancer hazard for man.

Question: Are there safe levels for human exposure to carcinogens; that is, are there threshold dosages below which we can be sure that no cancer will occur?

Answer: There is no adequate evidence for the existence of a safe threshold for any carcinogen. As far as we know, the frequency of tumor formation declines as dosage declines, but the risk of carcinogenesis may not disappear until the dosage reaches zero. Although high doses often are used in the tests, we must not assume that only high doses cause cancer. On the contrary, we must assume that low doses will also cause cancer, but at proportionally lower rates.

Human cancers have occurred following very low level exposure. Asbestos brought home on the clothing of asbestos workers, for example, has caused fatal cancer in members of the workers' families.

Question: Can we estimate the magnitude of the human cancer risk from the results of laboratory animal experiments?

Answer: Yes, we can estimate it in various ways, but the estimates usually are crude and we can't always depend on them. Animal tests tell us there is a hazard, but they don't tell us the extent of that hazard. Attempts to quantify the risk can lead to large errors, resulting from different methods of calculating risk, different susceptibilities between laboratory animals and humans, and potential interactions between the agent in question and other chemicals to which man is exposed.

Question: What happens when people are exposed to several carcinogens at the same time?

Answer: The resulting cancer rate may be higher than would be predicted by adding the risks from each carcinogen alone. Cigarette smoking and asbestos exposure, for example, are both carcinogenic, but asbestos workers who smoke are subject to a cancer risk that is far higher than would be expected by adding the risk from smoking to the risk from asbestos.

Question: What is the best way to prevent cancer?

Answer: Individually and collectively, we must make every effort to reduce or eliminate human exposure to carcinogens. This effort applies to agents known to cause cancer in humans, especially tobacco smoke. It also applies to agents that, on the basis of evidence gathered in animal or other experimental studies, are suspected of causing cancer in humans. Because everything does not cause cancer, such an approach to cancer prevention is reasonable and workable.

For further information

Cancer Information Services (CIS) can provide toll-free telephone answers to your questions about cancer cause, prevention, diagnosis and treatment.

Call 1-800-4-CANCER

In Alaska, call 1-800-638-6070 and in Hawaii, on Oahu, call 524-1234 (call collect from neighboring islands). Assistance for Spanish speakers is available during daytime hours in California, Florida, Georgia, Illinois, northern New Jersey, New York and Texas.

SECTION 3

The Growth of the Organic Farm and Food Movement
(with lists of retail and mail-order outlets)

The official U.S. Department of Agriculture definition of organic farming is producing food without pesticides, synthetically compounded fertilizers, growth regulators and livestock and poultry feed additives. This section contains articles describing the spreading awareness among farmers that going organic leads to both better quality and greater profitability. It is quite fascinating to read the specifics which explain just why. Consumers of course love the superior flavor, nutrition and safety of foods grown organically, and, as Marian Burros points out, the main-streaming of this trend is an outgrowth of this decade's health consciousness.

If you're at the point where you are asking yourself: "where can I find such food?" this is the part of Eating Clean with the address and telephone lists of retail stores and organic food mail-order suppliers compiled by Americans for Safe Food. Marian Burros sampled twelve companies form this list with "good to superb" results overall, except for lettuce which, she says, "does not travel well."

The Commercial Organic Farm:

On their 650-acre spread in central Ohio, the Spray brothers show that organic methods make good ecological – and economic – sense

BY JOHN THORNDIKE
COUNTRY JOURNAL, APRIL 1982, P. 39+.

On a sunny spring morning after a month of incessant rains, the soil in Ohio has finally dried out enough to be worked. Farmers all over the state have taken to the fields; a few are plowing, others are loading seed and fertilizer from their pickups into no-till corn planters, and most are pulling disk harrows over the rough furrows of land plowed earlier.

On the Martinsburg Road, south of Mt. Vernon, Glenn Spray and his brother Rex are out disking in the unwavering morning light, preparing their soil for field corn and soybeans. Glenn is driving the farm's largest tractor, a 1066 International with dual rear wheels and 120 horsepower, enough to pull a massive, 12-foot Miller disk weighing 3 1/2 tons. The polished round blades of the disk are set at opposing angles, and they churn through the earth but do not bury the top layer of the soil as a plow does. The earth wells up between the blades of the disk in an aerated wave, curling for an instant as if liquid. Moving at 5 1/2 miles per hour, 8 acres an hour, this disk is an awesome implement to anyone who has worked up ground by hand; 200 workers with mattocks and hoes could not chop and fluff up the soil as well, or as fast, as this tractor and disk.

With this level of technology and hundreds of acres to prepare in the midst of Ohio farm country, it doesn't seem at first glance that anything out of the ordinary is going on at the Spray farm. But the Spray brothers are part of a small minority of farmers who are quite at odds with mainstream United States agriculture: They raise their crops and livestock organically, and although they are disking their fields this morning in the same way as thousands of other Midwestern growers, many of their practices differ radically from those of farmers who rely on synthetic chemical fertilizers.

After a century of attrition, there remain some two million farmers in America, and of these the U.S. Department of Agriculture suggests that perhaps 2 percent farm organically. "Biological" and organic gardeners abound, but not full-scale organic farmers. A farmer who thinks of switching from synthetic chemical to biological methods of fertilizing and weed- and pest control is still apt to be given economic warnings and dire predictions by cooperative extension agents, bankers, and agricultural economists; yet for the past decade the Sprays have farmed their 650 acres without any synthetic fertilizers or pesticides, and have steadily produced healthy animals and better-than-average yields.

Glenn and Rex, both in their early fifties, work with Glenn's son Howard, and the three of them have a handsome way of making the job look easy. They get an enormous amount done without seeming to be in a hurry, and they are willing to stop in the middle of disking or cultivating to talk about their methods and plans. They are critical of so-called chemical, or conventional, farming, but they are without any traces of bitterness, and clearly they have reached their full maturity as farmers.

The Sprays' farm is about 220 acres larger than the current average for U.S. farms; 650 acres is greater than a square mile – an appreciable amount of farmland for three men to work. Most of it is bottom land with low rises and hills. It is open country with few hedgerows, but with woods of hickory, maple, and cherry in the distance. Glenn and Rex began farming together in 1957 and gradually expanded their holdings, buying the home farm when their father died in 1972. The Sprays have a number of mortgages to meet, and they work under the same economic pressures as most farmers. They farm as many acres as they do for the sake of efficiency, like other farmers who have expanded and survived into the eighties.

A sign on the barn announces that this is a demonstration farm for the production of organic food, but it simply looks like a prosperous family farm, one like many others. Glenn and Rex have ranch-style homes at either end of the farm, and there are barns, sheds, and corn storage bins. The Sprays have a dairy herd of forty cows, some fifty beef cattle, a warehouse for their fertilizer distribution business, and a great deal of farm equipment. Where this farm differs from others is in its soil.

For many years the Sprays have worked to conserve their soil and increase its fertility. That sounds quite reasonable, but in an age of chemical agriculture it is a marked anomaly for a farmer to have fertile soil. Fertile soil depends on a balanced pH level; on the presence of nitrogen, phosphorous, potassium, and other elements; and on an abundance of organic matter and microbial life in the soil to make these nutrients available to plants. Chemical fertilizers provide crops with water-soluble, instantly available nutrients, bypassing the slow, "natural" process in which nitrogen, phosphorus, potassium, and other plant requisites are broken down into a form that crops can use. Correctly applied, chemical fertilizers have helped many farmers produce championship yields, but while they do much for crops, they do little for the soil. Chemical farmers generally feed their plants well, but allow the soil slowly to exhaust its own organic matter and deteriorate in tilth. The concomitant use of harsh herbicides and insecticides, which kill off earthworms, beetles, and a host of microorganisms, can reduce the soil to little more than a hydroponic medium for which synthetic crop nutrients must be supplied each season. The loss of organic matter and microbial activity in the soil leads to soil compaction, erosion, and decreased resistance to pests, but many farmers have accepted these ills in return for the chemicals' ease of application, and are firmly, sometimes fanatically, convinced that no other way of farming is profitable.

The Spray brothers have farmed another way for the last ten years, using no herbicides, insecticides, or synthetic fertilizers. They have maintained consistently high crop yields and milk production by careful planning, some extra work, and some changes in technique. For example, they rotate their crops assiduously, and cultivate more than most farmers. They disk their fields but never plow them deeply, and they plant later in the spring than most farmers in order to work the ground more often and control more weeds before their corn and soybeans are seeded. They use a microbial fertilizer in place of the traditional nitrogen, phosphorous, and potash salts, and they spread all the manure they can.

One of the joys of looking at U.S. agriculture is that farmers – both organic and chemical – are a clever lot. Most of them have keen eyes for their land and notice details about it that even Thoreau would miss. From the enormous range of techniques, machinery, and possible rotations, each farmer invents his or her own system, with variables depending on the history of individual fields, the weather, the state of the soil, probable market prices, and so on. After talking to Glenn and Rex Spray, you realize that none of their moves in the field is by happenstance, but each choice is the result of years of observation and trial.

For example, consider their crop rotation. A good rotation does much to control weeds and disease and to improve fertility. Even the layman knows that crops should be rotated every year; yet there are fields in the Sacramento Valley that have been planted to tomatoes for a dozen straight years, just as potatoes are grown year after year on Long Island, and corn in Iowa. The practice of continuous corn growing has come under such attack that many Corn Belt farmers have switched to a two-year rotation of corn and soybeans, but the Sprays go this some better. Soybeans capture or fix nitrogen from the air, which helps with fertilization, but beans do not provide much organic matter for the soil. Plow-down crops of clovers and grasses do this, and for many years the Sprays have helped to maintain their soils by tilling under their hayfields.

The Sprays' rotation is corn, soybeans, wheat, and hay. In any given year different fields will be at different stages of the rotation, but stepping into the cycle at the end of the final year, the rotation follows this sequence:

The sod or hayfield mixture of grass and clover is disked under and left in rough condition during the winter, and as soon as the soil is dry enough to work in the spring, the field is disked again. At least two passes are made to kill off the sod and turn up the successive flushes of weed seedings. Corn is the first year's crop. A hybrid variety is seeded toward the middle of May and is weeded after it emerges by careful passes with a tractor and front-mounted cultivator, again more than once. Because the Sprays use no herbicides, they consider it essential to give row-cropped fields a total of four or five passes each spring to keep the weeds in check.

The corn is harvested in October or November, either by a combine which shells the corn or by a two-row corn picker which delivers ear corn. The field is disked, left during the winter, and then disked again in the spring, the same as the first year. Beans are seeded soon after the date for corn planting (when farmers talk "beans," they generally mean soybeans), preferably around the end of May. Beans are planted in rows, like corn, and weeds are controlled by the same row cultivators until the plants are too large for the tractor to pass safely along the rows. The beans are combined in late October, with a chopper mounted behind the combine to shred the vines and spread them over the field. When the nitrogen nodules have dried (the nodules are the mealy white clots on soybean roots that contain the atmospheric nitrogen fixed by the plant), the field is disked once more and wheat is planted immediately. At the same time as the wheat, timothy, a standard forage grass, is seeded. Both wheat and timothy get a start in the fall, and in the early spring of the next year a clover mixture, which germinates without the need for stirring the soil, is broadcast over the field.

The wheat grows up and its stalks dominate the field until it is harvested in July of the third year of the rotation. The combine removes the grain, and the stalks are clipped and baled to use for livestock bedding. (This straw is the only crop residue that is not directly returned to the soil, though it goes back eventually, enriched by cow manure and urine.) The timothy and clover now have light and room to grow, and if the year is wet enough, they may yield a first cutting of hay in August; or they will provide pasture for the animals.

In the final year a crop of hay is taken off the field in early June, with the second cutting in August, and the clover and grass are then fed to the soil by disking them in and letting them rot. This sacrifice of a good hayfield is one that many farmers have a hard time making, yet the Sprays are actually amending their current rotation to include a second year of wheat and

plow-down mixture, to guarantee that a steady stream of roots and top growth will enrich the soil as a green manure.

Organic matter and bacterial activity are the key to soil fertility, according to Glenn Spray. The amount of organic matter in the soil is often difficult to judge, even when walking in a field, but it is evident to farmers by the ease with which they can pull their disks or plows through the ground, and by how much water the soil will absorb in a heavy rain. Virgin mineral soils in the Midwest generally contained 4 to 5 per cent organic matter, but after a century and a half of farming, most of these soils now have less than 2 per cent organic matter, and many heavily farmed soils now contain less than 1 per cent. In common parlance, they have been corned to death. Without organic matter soil cannot hold moisture, and erosion on even faintly sloping lands becomes a major problem. Fields planted to corn are the most susceptible to erosion, with routine losses of ten to fifteen tons of soil per acre, every year. Extreme erosion losses have been documented: more than 28 tons of topsoil per acre in a single year from hilly cornfields.

Five years ago the Sprays tested for the amount of organic matter in some of their fields. One field had a respectable 3.6 per cent; others were reported at 4.8 and 5.4 per cent; and one field was analyzed at a remarkable 6.4 per cent organic matter.

"That is how we can absorb a six-inch rain with no run-off, if it doesn't come too fast," Glenn says. "But you can find plenty of fields with standing water after only a one-inch rain. There's nothing left in those soils to hold water, and that's what causes erosion."

Increased organic matter is also crucial for soil balance and fertility. In the last ten years the Sprays have added to their soil neither lime nor any major amounts of nitrogen, phosphorous, or potassium. They spread manure, but each year can only cover about 100 acres of the farm's 650.

"Manure is important to the soil," Rex says, "but it's not essential. Green manures will do the same job for you when you plow down clovers and grass. There are plenty of organic farmers who are exclusively grain farmers and who don't have the manure to spread."

The fertility of the Sprays' soil is also based on another factor that is hard to see, even when holding the dirt in one's hands: the abundance of soil organisms. The Sprays assert that small soil animals and microflora buffer the soil pH and make nutrients available to the plants.

"Healthy soils should have an abundance of fish-worms," Glenn says. The presence of earthworms (Lumbricus terrestris, what Glenn's calling a fish-worm) is a general sign that other, smaller organisms are also prospering in the soil. Some of these are visible to the naked eye: beetles, centipedes, millipedes, springtails, and spiders. Others are microscopic: bacteria, fungi, actinomycetes, and protozoa. This web of living organisms has an intimate effect on the soil's chemical make-up, and all organisms thrive best when the soil contains abundant organic matter.

To boost the microorganisms in their soil, the Spray brothers use a microbial fertilizer called Wonder Life, formulated by a Des Moines, Iowa, firm; and they supplement their farm income slightly by distributing Wonder Life's Special N fertilizer and livestock products. At the Spray farm, fertilizer usually goes on the fields before corn is planted, in the first year of the Sprays' crop rotation sequence. The entire Wonder Life program includes recommendations on tillage, rotation, and fertilizers, and Wonder Life may include in its fold a full third of the country's large-scale organic farmers. Wonder Life's "humate" fertilizer uses a base of lignite, an inferior coal that contains humic acids, the chemicals that allow soil humus to hold water and that aid in the breakdown of the soil's naturally occurring phosphorous and potassium. The fertilizer also provides a small amount of calcium, magnesium, and sulfur, numerous trace elements, and some forty strains of beneficial microbes. Wonder Life does not identify the microorganisms in its fertilizer, and the company does not make exact claims for the fertilizer's effectiveness; instead the company points to the success of its fertilizer's users and add, "Our product is only a fourth of our program."

Farmers who rely on synthetic chemicals look right past such fertilizers, because for each year's crop they need soluble, instantly available forms of nitrogen, phosphorus, and potassium for their depleted soils. Microbial fertilizers (called soil conditioners by many) depend on the soil's having a certain receptivity – that is, these fertilizers depend on a sufficient level of organic matter being present in the soil so the microorganisms they supply can prosper.

In the agricultural mainstream, fertilizers are judged almost exclusively by yields. The Sprays are more concerned with the overall fertility of their soil than with specific yields, but farming is not a mystical hobby to them, it is a business with mortgage notes and fuel bills to pay. In 1980 their corn yielded more than 125 bushels per acre, their soybeans averaged 48 bushels, and their wheat 45. (These compare with Knox County averages of 111 bushels for corn, 34 for beans, and 39 for wheat.) In 1979 the Sprays had an official yield check run by the county, in which two fields yielded more than 140 bushels of corn per acre and one yielded 180 bushels. The Sprays' fertilization bill this year on their cornfields was less than $50 an acre, and that application is made only once in their four-year rotation; a conventional grower who followed corn with corn would have spent close to $300 per acre on fertilizer over the same period.

The USDA's milestone 1980 report on organic farming went to some lengths to explain that large-scale organic growers do not, as a rule, advocate a return to the farm technology of the 1930s. The Spray Farm is technologically similar to most large U.S. operations and is run with the help of a wide array of implements and modern equipment. Not all the machinery is brand-new, of course, but the Sprays own and operate all of the following: eight International tractors, a New Holland self-propelled combine with separate grain and corn heads, a Chevrolet 2 1/2-ton truck, a four-row planter for corn and beans, a grain drill, a two-row corn picker, two front-end loaders, three manure spreaders, two row cultivators, two smoothing harrows, a rotary mower, a portable hammer mill for grinding feed, a rotary hoe for field preparation, two grain auger-elevators and a hay elevator, two offset tandem disks, a sickle bar mower, a mower-conditioner for hay, a hay rake, a baler, four flat-bed wagons, four gravity wagons for hauling grain, and a pipeline milking system and bulk tank.

Eight tractors, one asks, isn't that an extravagance? There are times when they use them all, or nearly so. In late June of 1981, for example, after the wettest Ohio spring in decades, dry weather finally allowed the Sprays to seed their last soybeans of the year. As a teenager rotary-hoed the field ahead of him, Rex seeded the beans with a second tractor. Across the road Glenn was in the corn, giving a first pass with the row cultivator, and his son Howard was pulling in wagonloads of hay from another field, some of the first of 10,000 bales. A fifth tractor pulled the baler, and a sixth powered the elevator that Howard and a third Spray brother were using to send bales up into the barn for storage. The final two tractors were in reserve in case of a breakdown, or ready to be mounted with the hay rake and mower.

Few of these farm implements represent the latest in technology, and there are no tractor cabs here with air conditioning and tape decks. To run a large farm requires a great deal of machinery, however, and the Sprays are representative of the growing number of farmers who have changed their farming philosophies but kept their hardware.

Notably absent from the Sprays' list of farm implements is the moldboard plow. They don't own one and haven't deep-plowed their land for more than ten years. A moldboard – still the standard American plow – shears a layer of earth and folds it over the previously plowed furrow, burying organic matter and surface residue too deep to allow for rapid decomposition. The Sprays do not use a chisel plow either, although an increasing number of organic farmers do. A chisel plow looks something like an oversized cultivator, with thin shanks that go down 12 or even 24 inches.

Because much of the Sprays' land is drained by clay tiles buried by hand to a depth of only a foot or two, a chisel plow would damage their drainage system. They rely instead on deep-feeding grasses and weeds to prevent any formation of hardpan and to bring up minerals from the subsoil.

After a decade of organic practice the Sprays are sure of their program. The farm's animals are healthy; corn and soybean yields are higher than both county and state averages; and erosion has been held to a minimum. Two serious pests in Ohio, the corn borer and the corn rootworm, are fought in a pitched battle by chemical farmers, but the Sprays use no pesticides at all and suffer little damage. Yet the Sprays still have difficulty gaining acceptance for their program from the Knox County Extension Service, agricultural cooperative board members, and the other local farmers.

"The county agency will come out and look over what we're doing," Rex says. "He'll say, 'You're doing a fine job here, but this isn't for everybody. You've got to be a good manager to farm this way.'"

The Sprays believe they have learned some important lessons about farming and want to share them, but it is slow going; it takes many explanations and farm demonstrations to persuade a few farmers about organic techniques. Glenn himself recalls that as a young man he thought that the author Louis Bromfield "seemed like a nut. He had a funny reputation; he was an advocate of not plowing." Bromfield's Malabar Farm, only thirty miles from Mt. Vernon, was a major center for organic farming and research in the late forties and early fifties.

Might some farmers think today that the Spray brothers are a little nuts?

"They might," Glenn laughs, "they just might. Especially close to home. You can be an expert far from home, but not close by."

The first few years of switching to organic methods are still regarded as the most difficult, especially if the soil has been heavily treated with chemical fertilizers and pesticides. Yet for the Sprays, that transition was not a traumatic period. Whereas most farmers anticipate a sharp decrease in yields, Rex only says laconically, "When we went off chemicals, we didn't notice an increase in production right away." The Sprays probably had an easier time of it because their soil had previously been treated well; they and their father, Herbert Spray, used chemical fertilizers as well as Atrazine and 2,4-D, but rarely any of the other chlorinated hydrocarbon or organophosphate pesticides such as Aldrin, Dieldrin, Parathion, and Malathion. They rotated their crops, chopped up plant residues, and spread manure and bedding from the dairy barn, so their organic matter was 3 to 4 per cent even in their first year of "biological" farming. Glenn admits that not all farmers would have such an easy time of it, and some might be advised to phase into organic methods step by step.

When the Sprays made the transition to organic farming, it was not because they worried about the health hazards of chemicals ("Though we would now," Glenn says, "because the chemicals are so much stronger"), but because chemicals were no longer doing the job for them. The immediate problem was an inability to control grasses in their fields; they subdued outbreaks of quack grass only to be infested by fall panicum. Glenn says, "We figured that if we killed one off, then something else would take its place that would be just as bad or worse." So in 1970 they stopped using herbicides and began to cultivate more, which has given them cleaner fields. There are still some weeds, of course; some they live with, and occasionally they weed their corn and soybean fields by hand to stop the jimsonweed from going to seed: "We walk the fields sometimes on that one," Glenn says. "Those rascals can get as big as trees."

Farming organically hasn't solved all the Sprays' problems. They still fight vagaries of the weather, the unending resurgence of weeds, occasional insect damage, and peak seasons when there is too much to do – the same as other farmers. There is nothing pat and resolved about farming, which remains a testing ground for the clever. During the last forty years, most farmers have felt that the smart money has been with synthetic chemicals, but the Sprays think it's a mistake to consider only yields. They believe a good farmer looks after the soil, first and last. Ten more years might tell us definitively which farmers have been the cleverest, or the debate may still go on, but the Spray farm is proof that large-scale organic farming can be successful, today.

Organic Farming is Blossoming:

Pollution And Costs Turn Growers Away From Use of Chemical

BY WARD SINCLAIR

THE WASHINGTON POST, MONDAY, NOVEMBER 23, 1987

When Wisconsin Agriculture Secretary Howard Richards learned that his state was entitled to millions of dollars from the penalties levied on oil companies for overcharges in the 1970s, he set wheels turning in behalf of an idea heavy with political irony.

Why not take some of that money, he wondered, and put it to work to help wean Wisconsin framers away from the expensive petroleum-based fertilizers and pesticides that increasingly were eroding their profits and threatening the state's environment?

When the penalty funds came through from Washington earlier this year, the state earmarked $2 million for a project intended to inspire farmers to take up farming techniques that will save more and pollute less.

Applications are rolling in. And project director Ken Rineer estimates that about $800,000 will be handed out in the next year to bankroll the most ambitious state-run "sustainable agriculture" project in the country.

"Secretary Richards' interest is in keeping farmers on the farm. The profitability angle appeals to him," Rineer said. "But part of the reason also is environmental. We have a problem with groundwater pollution, and perhaps we can change that if we can show farmers methods for reducing contamination."

Similar, if less dramatic, things are happening coast to coast. Pushed by farmers' concerns about costs, environmental pollution and personal health, state governments and universities are gradually climbing on the bandwagon of change in agriculture.

Whether it is called sustainable, renewable, regenerative, organic, low-input or something else, it is essentially the same: an attempt to reduce or eliminate chemical pesticides and to improve soil fertility by returning to the crop rotations that were common before the age of chemical fertilizers.

The University of Maine now offers an undergraduate degree course in sustainable agriculture. The University of California at Santa Cruz has become a center of sustainable agriculture teaching. Vermont, Minnesota and many other universities include sustainable courses in their curricula. Iowa State University is setting up a major research center for sustainable farming. Others have begun research programs.

"There's almost an explosion of interest. It's hard to track–and I'm doing this full-time," said Garth Youngberg, executive director of the Institute for Alternative Agriculture in Greenbelt. "The real explosion is occurring among conventional farmers. It is a convergence of economic stress, the search for more viable alternatives and concern about environmental matters, particularly water quality.

"There is anxiety about how long the government-support programs will last, anxiety about how many farmers could survive without the programs. They are positioning themselves. There's a lot of re-thinking, searching and looking going on...a sense that they have to do something while there's time."

Warren Sahs, a University of Nebraska agronomist who championed organic-farming research years before it became popular, added: "There's a definite groundswell among the land-grant universities, an upsurge of interest in the last two or three years, especially among younger, new staff members....There's a bandwagon going, and a lot of people are jumping on. I'm not being critical; it's a wonderful development."

But here in Washington, the bandwagon described by Sahs is stuck in a rut somewhere between Capitol Hill and the U.S. Department of Agriculture while USDA bureaucrats apparently wrangle over control of the first congressional spending ever earmarked for government research into "low-input" farming methods.

After four years of opposition, USDA agreed to language in the 1985 farm bill that directed research into low-input techniques in the interest of reducing farmers' costs. Although USDA sought no funding for the studies, it agreed this year to a $7.5 million appropriation sought by a coalition of alternative-agriculture groups.

The Senate's agricultural appropriations bill for fiscal 1988 designated about $7 million for low-input farming studies, with an emphasis on getting new information to farmers. But House appropriators limited new spending to $2.6 million, reacting to Agricultural Research Service claims that it already spends about $100 million annually on research with low-input applications.

While official Washington debates, farmers such as Carl Pulvermacher of Lone Rock, Wisconsin, are not waiting for the experts' help. It is almost a case of vice versa; about a dozen university scientists and county extension agents showed up earlier this month when Pulvermacher and fellow no-chemical farmers held a field day to demonstrate their work.

Pulvermacher, a former chemical fertilizer salesman, has raised corn and alfalfa without chemicals of any sort since he and his wife bought the land they were renting in 1982. His dairy herd's production is far above the state average. His corn and alfalfa, fertilized with manure, produce exceptional yields. Crop rotations control weeds and pests. He calculates that his costs are at least $5,000 less per year than neighboring farmers.

"The reason I got into this was an idealist environmental reason. The reason I stay in it is profitability," he said. "I maintain that sustainable agriculture has kept me on this farm . . . Some people say I'm going back 40 years, but what I am doing is just common-sense agriculture."

But, he added, textbooks are not likely to give farmers easy answers. "A lot of my decision-making is seat-of-the-pants; I can't sit down in March and make my planting decisions. There's no recipe for this, and it bothers people. A lot of farmers want a recipe, and it isn't there," Pulvermacher said.

"I was forced to make the change because of economics," he added. "If I didn't have the interest and principal costs on this farm, holy fuzz, would I be having a riot. There's nothing radical about using less inputs."

The Fresh Appeal Of Foods Grown Organically

BY MARIAN BURROS
THE NEW YORK TIMES, WEDNESDAY, JANUARY 28, 1987

If any one event focused public attention on the hazards of chemical contaminants in food, it was the poisoning in 1985 of thousands of watermelons with the pesticide aldicarb. "It seemed to crystalize consumer and producer awareness, because there were so many people who got so acutely sick," said Mark Lipson, executive secretary of California Certified Organic Farmers.

Aldicarb in watermelons was preceded by aldicarb in potatoes and ethylene dibromide, or EDB, in grains and citrus. It was followed by daminozide in apples and heptachlor in milk. Episodes such as these, and their attendant publicity, have fueled a significant increase in demand for organically raised food, free of chemical pesticides, herbicides, fungicides and fertilizers, and for "natural" beef, veal and poultry—animals raised without hormones or antibiotics.

"The attitude of the 1960s has been mainstreamed," said Judith Gillan of Belchertown, Massachusetts, the secretary of the Organic Food Production Association of North America. "You don't have to be out in the woods with your hair down your back and refuse to talk to people who go to Wall Street every day to care about organically grown food."

Once synonymous with hippies, health faddists and environmentalists, organically grown food is now sought out by an ever-increasing proportion of mainstream America, people who care about freshness and taste as well as cleanliness and safety. "They include mothers concerned about what they feed their children and older people who want food that tastes like it used to taste," said Joseph Dunsmore, president of Organic Farms Inc. in Beltsville, MD., the largest distributor of organically grown food on the East Coast.

Unfortunately, not all food labeled "organically grown" actually is, and it is almost impossible for consumers to distinguish between the genuine and the fraudulent by looking. The industry is working on a verification system that will certify producers who adhere to certain principles of organic farming. Until then, the more consumers know about acceptable standards for the production and sale of organically grown food, the better able they will be to make informed choices.

The search for uncontaminated food is primarily the outgrowth of the new emphasis on a healthy diet. "It's all part of the health consciousness," says Deborah Schechter, the director of Americans for Safe Food in Washington, a coalition of consumer, environmental and farm groups.

Organic farming is estimated at between one and 1.5 percent of total agricultural production in the United States. There are no statistics to document its popularity over the last few years, however, only individual distributors, growers and processors, who have witnessed an upsurge in business.

When Organic Farms opened six years ago, it had 20 farmers; now it has 400. Five years ago it was selling to 10 stores in the New York area; it currently sells to 50. Shoppers could find only four organic farmers at New York City's Greenmarkets five years ago; today, 15 farmers sell contaminant-free produce. California Certified Organic Farmers had 150 members two years ago; there are 300 now. Meanwhile, the Organic Foods Production Association, which began two years ago, now represents 3,000 to 5,000 people.

Perhaps the clearest indication that organically raised food has gained popularity is its presence in conventional supermarkets. Grand Union sells Pine Ridge chickens and Coleman's "natural" beef, raised without antibiotics and hormones. Bread and Circus, a chain of four stores in Boston, features organically grown produce. In addition, several supermarket chains, including Grand Union and A.& P., have refused to sell apples sprayed with daminozide, and for its baby food, H.J. Heinz will not accept food treated with 12 chemicals the Environmental Protection Agency has labeled potentially hazardous.

The organic food business has become large enough for the industry to worry about preserving its integrity. Only eight states have laws governing what can be labeled "organic," but New Jersey and Connecticut are not among them. Legislation is being drafted in the New York State Legislature by Senator Nancy Lorraine Hoffman, Democrat of Syracuse.

The basic principles governing the cultivation, distribution and labeling of organically grown foods–principles agreed to by legitimate industry members–include the following:

• Production, harvest, distribution, storage, processing and packaging cannot be performed with the use of synthetically compounded fertilizers, pesticides or growth regulators.

• Meat, poultry or fish must be produced without chemicals or drugs to stimulate or regulate growth or tenderness. In some definitions feed and water must be free of pesticides or other synthetic chemicals.

• Anyone selling, labelling or advertising organically grown products must maintain records attesting to proper farming techniques.

Between 25 and 40 organizations can provide verification that organic food meets certain standards. But the standards differ. So, the Organic Foods Production Association is seeking national guidelines for the verification process. "Any farmers group that gets together and verifies itself is unacceptable," said Tom Harding, a farmer in Windgap, Pennsylvania, who is the association's president. "We want third-party,

independent verification." Still, some highly regarded organic food companies, such as Walnut Acres in Penns Creek, Pennsylvania, certify their own products.

Five methods are used in the verification process: questionnaires, affidavits, on-site inspection by third parties, laboratory analyses and audits. Legitimate vendors of organically grown food should have such certification and should be willing to make it available to customers. Consumers, meanwhile, should not hesitate to ask for certification. According to Ms. Gillan, independent certification will become an important selling tool. "All of us in the business who want to protect organic know that fraudulent operators will do us in," she said.

Many farmers, suffering the effects of a depressed farm economy and fearing constant exposure to toxic chemicals, are also seeking organic alternatives.

"The Agricultural Research Service Experiment Stations are looking into the use of organic farming," said Agriculture Secretary Richard E. Lyng, "and the newly created office of small-scale farming at USDA answers questions and makes policy regarding organic farming."

As organic farming has become more widespread, its products have become better looking than they were 20 years ago. They have to be. "No one is going to make a pilgrimage to the organic produce section to buy poor-quality stuff," Ms. Gillan said.

"We've gotten produce quality up 10 times better than it was even five years ago," Mr. Dunsmore said. "We are using more sophisticated refrigeration, temperature and humidity control, and we are able to supply better quality, which customers have responded to."

Customers, however, must be prepared to spend more. Organic farmers are paid on the basis of production costs; conventional farmers are not. "We pay our framers enough money so that they can continue to be there next year," said Paul Schwartz of the Little Bear Trading Company in Winona, Mississippi, which processes organically grown food. "Obviously, that is self-serving. But if there are no farmers, we can't go out on the market to buy the next year."

Some retailers also take stiff mark-ups. While conventionally grown broccoli was generally selling for $1.39 a pound last seek in Manhattan supermarkets, organically grown broccoli was as much as $1.89 a pound.

But more and more people are willing to pay the difference. Despite the expense, "People are beginning to make the connection between high-quality food and organic production," said Mr. Lipson of the California organic farmers' group. "That is really a turnaround from what the general image was a few years ago." *See list on next page.*

The Fresh Appeal Of Foods

For The Buyer, Picking Taste Over Looks

Organically grown foods are usually labeled as such. Whether or not they actually are organic is another matter; consumers should ask to see verification.

Beyond that, consumers must trust their merchants. They also must understand that such foods will not look like the conventional variety because they are not treated with chemicals. Apples won't shine; some produce may be scarred or mottled. Organically grown foods spoil fairly quickly, because they have not been waxed or chemically preserved.

The tradeoff is that these foods generally evidence a remarkable depth of flavor and richness–particularly since merchants have finally learned to handle perishable foods correctly.

Finding items certified organic by an independent group can be difficult. Americans for Safe Food, a coalition of consumer, environmental and farm organizations, has compiled a list of mail-order sources, which is available by sending a stamped, self-addressed business-size envelope to Organic Mail Order, Americans for Safe Food, P.O. Box 66300, Washington, D.C. 20035.

With rare exceptions, these companies accept only money orders and personal checks, or they will ship c.o.d. Shipping is extra.

Certified organic food sampled from 12 companies on the Americans for Safe Food list ranged from good to superb, with only one or two disappointments. (Lettuce, for example, does not travel well.) The comments below apply only to the foods sampled and not to the entire range of products offered. In addition, eight stores in New York City, New Jersey and Connecticut were visited.

Be Wise Ranch, 9018 Artesian Road, San Diego, California 92127; 619-756-4851: limes, lemons, avocados and oranges.
Navel oranges, 40 cents a pound, were superbly sweet.

Cascadian Farm, Star Route, Rockport, Washington 98283; 206-853-8175: fruit preserves sweetened with honey; dill pickles, regular and low-sodium.
The preserves were acceptable. The low-sodium pickles were the best of their kind; six 10-ounce jars, $14.25.

Country Life Natural Foods, Oakhaven, Pullman, Michigan 49450; 616-236-5011: beans, grains, seeds, nuts and dried fruits.

The brown rice, 57 cents a pound, tasted like conventional rice. The prunes, $1.59 a pound, were the biggest, fattest and sweetest ever tasted. Walnuts, $2.69 a pound, were free of the bitterness walnuts often have.

Deer Valley Farm, R.D. 1, Guilford, New York 13780; 607-764-8556: chicken, beef, veal, pork, fruit, vegetables, grains, herbs, juices, pasta, oils, soups, spreads, seasonings, baked goods, confections and nuts.
Of the items sampled, the chicken, $2.19 a pound, was extraordinary– white in color with very little fat and very flavorful. Veal loin chops, $4.89 a pound, were tender. Pork chops, $3.35 a pound, were quite fatty but very flavorful. Beef, $2.75 a pound, was virtually tasteless. Breads, ranging in price from $1.19 to $1.49 a pound, were slightly stale but toasted nicely and had a good, nutty flavor. The sunflower cake, 12 ounces for $2.29, and strudel, $2.50 a pound, would appeal only to those who eat to live.

Ecology Sound Farms, 42126 Road 168, Orosi, California 93647; 209-528-3816: kiwis, persimmons, Asian pears, plums, oranges, garlic and dried fruit.
Kiwis and persimmons were sampled, with delightful results. Kiwis, $15 a flat (36 pieces, or 8 pounds), were excellent; persimmons, the Fuyu variety at $14 a flat (28 pieces, or 10 pounds), were firm but sweet.

Living Tree Center, P.O. Box 797, Bolinas, California 94924, 415-868-2224: almonds and almond butter, garlic and strawberry jicama.
The almond butter, $6.65 a jar, was full of nutty flavor and a delightful substitute for peanut butter. The strawberry jicama, $1.75 a pound, was crisp and crunchy.

Oak Manor Farms, Tavistock, Ontario, Canada NOB 2RO); 519-662-2385: flours, grains, beans, seeds, cereals and coffee.
The 15-grain cereal, $2.88 per kilo (2.2 pounds), was outstanding. The whole-wheat cream of wheat, $1.94 per kilo, was delicious. For speeding cooking do not follow cooking instructions. Instead, combine grains and water, and cook until the water has been absorbed.

Pine Ridge Farms, P.O.Box 98, Subiaco, Arkansas 72865; 501-934-4565: chicken and turkey, shipped frozen every other Monday.

Chicken was tender and juicy with a deep, rich flavor. Light yellow in color. A 21-pound package, including three whole chickens, costs $49.

Rising Sun, P.O. Box 627, Milesburg, Pennsylvania 16853; 814-355-9850: beef, poultry, lamb, pork, fruits, vegetables, beans, seeds and grains.

Most of the produce sampled was exceptional, especially the broccoli, $1.50 a pound; beets, $1.50 to $1.65 a pound, and winter squashes, 75 cents to 85 cents a pound. All were full of flavor and delicious served without seasoning. The exception was red leaf lettuce, which had begun to rust.

Smile Herb Shop, 4908 Berwyn Road, College Park, Maryland 20740; 301-474-4288: fruits and vegetables.

Perhaps because of its proximity, produce from this shop arrived in the best condition of the mail-order produce sampled. The Granny Smith apples, $1.21 a pound, were crisp, tart and sweet. The citrus fruit was richly flavored: navel oranges, 95 cents a pound, and royal mandarins, 75 cents a pound. Broccoli was sparkling green and flavorful; $2.65 bunch (about one and a half pounds).

Sun Mountain Research Center, 35751 Oak Springs Drive, Tollhouse, California 93667; 209-855-3710: dried herbs.

Rosemary, thyme, oregano, savory and basil were redolent with fragrance; $1.50 for 1.5 ounces.

West Valley Produce Company, 726 South Mateo Street, Los Angeles, California 90021; 213-627-4131: fruits and vegetables.

Winter squashes, 50 cents a pound, potatoes, 40 to 50 cents a pound; broccoli, $1.15 a pound; cauliflower, $1 a head, and avocados, 80 cents a pound, all arrived in fine condition and were excellent. Two varieties of lettuce did not travel well and arrived with rust and decomposing outer leaves.

A number of shops in the New York City area sell certified organically grown foods. The following were visited (with one exception), but the food was not sampled. The prices of several vegetables are cited for comparison.

NEW YORK CITY

Back to the Land, 144 Seventh Avenue (Carroll Street), Brooklyn, 718-768-5654.

A small selection of nice-looking produce in a neatly maintained store. Broccoli, $1.85 a bunch; carrots, 79 cents a pound.

Commodities, 117 Hudson Street (North Moore Street), 212-334-8330.

A well-kept store stocking a large variety of produce. Except for celery and lettuce, the food looks very good. Broccoli, $1.19 a pound; carrots, 79 cents a pound.

Integral Yoga, 250 West 14th Street, 212-243-2642.

The few nonorganic items are so marked. Produce looks very good, and the selection is extensive. Broccoli, $1.89 a pound; carrots, 52 cents a pound.

Prana Foods, 125 First Avenue (St. Marks Place), 212-982-7306.

The assortment is very small and the store not especially tidy, but the food looks good. Broccoli, $1.59 a pound; carrots, 70 cents a pound.

Whole Foods, 117 Prince Street (Wooster Street), 212-673-5388.

A well-stocked store. Produce is well tended and appealing. Red or green cabbage, 89 cents a pound; carrots, 79 cents a pound, and fresh Pine Ridge chickens, $2.49 a pound.

NEW JERSEY

The Third Day, 220 Park Avenue, East Rutherford, New Jersey; 201-935-4045.

On a Monday morning, the store was low on produce but what was there, with the exception of the apples, looked good. Romaine lettuce, 99 cents a pound; carrots, 69 cents a pound.

Hoboken Farm Boy, 128 Washington Street, Hoboken, New Jersey; 201-656-0581.

A large selection of appealing produce. Mineola tangelos, three for $1; winter squash, $1.19 a pound.

CONNECTICUT

Fountain of Youth, 1789 Post Road East, Westport, Connecticut; 203-259-9378.

This well-maintained store carries a wide variety of items and a limited supply of good-looking produce. Broccoli, $1.95 a pound; carrots, 59 cents a pound.

LONG ISLAND

Rising Tide Natural Market, 42 Forest Avenue, Glen Cove, Long Island; 516-676-7895.

The store was not visited, but it has a reputation for a wide selection of organically grown as well as conventional produce.

Americans For Safe Food
Sources of Organic Food in the District of Columbia Area

We suggest calling these sources to determine the availability of organic items.

Beautiful Day
5010 Berwyn Road
College Park, MD 20710
(301) 345-6655

The Bethesda Co-op
7945 MacArthur Blvd.
Cabin John, MD 20818
(301) 320-2530

Cash Grocer Natural Foods
1315 King Street
Alexandria, VA 22301
(703) 549-9544

Glut Food Co-op
4005 34th Street
Mt. Rainier, MD 20712
(301) 779-1978

Gairsan Natural Foods
1368 H Street, NE
Washington, D.C. 20002
(202) 396-0034

The Good Food Store
1825 Columbia Road, NW
Washington, D.C. 20009
(202) 462-5150

Healthway Natural Foods
7037 Columbia Pike
Arlington, VA 22003
(703) 354-7782

Hugo's Natural Food Market
3813 Livingston Street, NW
Washington, D.C. 20015
(202) 966-6103

Kennedy's Natural Foods
1015 West Broad Street
Falls Church, VA 22045
(703) 533-8484

Low Sodium Market
4901 Auburn Avenue
Bethesda, MD 20814
(301) 657-3033

Montgomery County Market
804 Baltimore Road
Rockville, MD 20815
(301) 424-0787

Natural Foods Supermarket
6184-A Arlington Blvd.
Falls Church, VA 22044
(703) 536-2010

Naturally Yours
2120 P Street, NW
Washington, D.C. 20037
(202) 429-1718

Organic Farms
10714 Hanna Street
Beltsville, MD 20705
(301) 595-5151

Organically Yours
6441 Sienna Court
Falls Church, VA 22043
(703) 241-4083

Perfect Health Food Club
c/o Karen Groover
7801 12th Street, NW
Washington, D.C. 20012
(202) 829-6867

Takoma Park Co-op
623 Sligo Avenue
Silver Spring, MD 20910
(301) 588-6093

Tuscarora Valley Beef Farm
P.O. Box 15839
Chevy Chase, MD 20815
(202) 588-5220

Wheaton Health Foods
2656 University Blvd. West
Wheaton, MD 20902
(301) 933-3066

Women's Community Bakery
736 7th Street, SE
Washington, D.C. 20003
(202) 546-7944

Yes! Natural Gourmet
1015 Wisconsin Avenue, NW
Washington, D.C. 20007
(202) 338-1700

Organic food can also be ordered via mail-order companies. Americans for Safe Food has published a list of these sources. For a copy, send a self-addressed, stamped envelope to Americans for Safe Food, P.O. Box 66300, Washington, D.C. 20035.

AMERICANS FOR SAFE FOOD

Mail Order Guide to Buying Contaminant-Free Foods

Tired of pesticide-and drug-tainted foods? Well, help may be as close as your mailbox. The Center for Science in the Public Interest (CSPI) has published a list of mail-order sources of contaminant-free food, including Alar-free apples; antibiotic- and hormone-free beef, chicken, lamb, turkey, eggs, and cheese; and pesticide-free grains, fruits, vegetables, soup, pasta, and baked goods.

"It's gradually becoming easier for consumers concerned about potentially dangerous chemicals in their food to obtain safe alternatives," said Deborah Schechter, director of Americans for Safe Food (ASF). ASF, a national coalition of consumer, environmental, and farm organizations spearheaded by CSPI, is working to increase the availability of safer food.

"Our goal is for consumers to be able to purchase safe food conveniently at their local grocery stores," Schechter said. "Until then, many organic farmers will ship everything from soup to nuts right to your door."

According to Schechter, "In addition to providing more healthful food, these mail order sources enable the consumer to support conscientious farmers who do not pollute the environment; do not endanger their workers, themselves, and their families; and do not cram animals into such cramped quarters that they need drugs to survive."

The List

The following grocers and distributors can ship their products directly, usually via UPS, and do not require a minimum order unless otherwise noted. Write or call them for a complete listing of products and prices.

To minimize shipping costs, order in large quantities from a firm close to home. You may want to visit local farms. You can also call to find out which retail stores in your area carry their products.

We have noted below whether products are certified organic. A "(C)" indicates that all products listed are certified organic, which means that an independent agency verifies that farming practices conform with a prescribed set of standards. Those suppliers that establish their own standards are considered "self-certified", and are labeled with an "(S)". An "(N)" indicates that the products are not certified. For more information about certification programs, Americans for Safe Food has a fact sheet available upon request.

A listing here does not constitute an endorsement by Americans for Safe Food or the Center for Science in the Public Interest. CSPI cannot guarantee the quality or supply of these companies' products.

ARIZONA
Arjoy Acres
HCR Box 1410
Payson, AZ 85541
(602) 474-1224
Garlic, elephant garlic, shallots. (N)

ARKANSAS
Mountain Ark Trading Company
120 South East Avenue, Fayetteville, AR 72701
(800) 643-8909
Vegetables, miso, seasonings, rice, pasta, fruit, spreads, oils, beans, soup. Certification varies; products labeled accordingly.

Pine Ridge Farms
P.O. Box 98
Subiaco, AR 72865
(501) 934-4565
Chicken, turkey, grass-fed beef. Only poultry is certified.

CALIFORNIA
Ecology Sound Farms
42126 Road 168
Orosi, CA 93647
(209) 528-3816
Kiwi fruit, persimmons, Asian pears, plums, oranges, garlic, dried fruit. (C)

Giusto's Specialty Foods, Inc.
241 East Harris Ave, S.
San Francisco, CA 94080
(415) 873-6566
Breads, cakes, grains, spices, cereals, flour, oil, seeds, cookies, yeast. (N)

Jaffe Bros.
P.O. Box 636
Valley Center, CA 92082-0636
(619) 749-1133
Dried fruit, nuts, brown rice, pasta, oil, grains, cereal, beans, peas, seeds, olives, candy. (N)

Lee Anderson's Covalda Date Co.
P.O. Box 908
Coachella, CA 92236
(619) 398-3441
(Mon-Fri 9-12 & 1-4:30)
Dates, pecans, citrus, figs, raisins. (S)

Living Tree Centre
P.O. Box 797
Bolinas, CA 94924
(415) 868-2224
Almonds, almond butter, silver-skin garlic, Red Jerusalem artichokes, various vegetable seeds. Only artichokes and seeds are certified.

Timber Crest Farms
4791 Dry Creek Road
Healdsburg, CA 95448
(707) 433-8251
A wide variety of dried fruits. (N)

Joe Soghomonian
8624 S. Chestnut
Fresno, CA 93725
(209) 834-2772
Grapes and walnuts in season only, raisins year-round. (C)

West Valley Produce Co.
726 South Mateo Street
Los Angeles, CA 90021
(213) 627-4131 or 629-1656
Fruits, vegetables. Certification varies; products labeled accordingly.

Be Wise Ranch
Bill Brammar
9018 Artesian Road
San Diego, CA 92127
(619) 756-4851
Limes, lemons, avocados, oranges. A minimum order of 40 lbs. of citrus and 25 lbs. of avocados is required. (C)

Sun Mountain Research Center
35751 Oak Springs Drive
Tollhouse, CA 93667
(209) 855-3710
Herbs, herb seeds, (C)

G & J Farms
Gregory F. Gaffney
4218 W. Muscat
Fresno, CA 93706
(209) 268-2835
Apricots, peaches, assorted vegetables. (C)

CONNECTICUT
Butterbrooke Farm
78 Barry Road
Oxford, CT 06483
(203) 888-2000
75 varieties of chemically untreated, open-pollinated, short maturity

seeds. (N)

IDAHO
Brown Company
P.O. Box 69
Tetonia, ID 83452
(208) 456-2500 or 456-2629
Idaho potatoes, seed potatoes. (N)

IOWA
Frontier Cooperative Herbs
P.O. Box 299
Norway, IA 52318
(319) 227-7991
Herbs, spices. (S)

MAINE
Fiddler's Green Farm
R.R. 1, Box 656
Belfast, ME 04915
(207) 338-3568
Pancake, muffin, and spice cake mixes; breakfast gift pack including pancake & muffin mix, maple syrup, honey. (N)

MARYLAND
Smile Herb Shop
4908 Berwyn Road
College Park, MD 20740
(301) 474-4288 or 474-8495
Fruits, vegetables. Certification varies; products labeled accordingly.

Tuscarora Valley Beef Farm
P.O. Box 15839
Chevy Chase, MD 20815
(301) 588-5220
Lamb, veal, beef, nitrite-free bacon, lower-fat chicken sausages. (N)

MICHIGAN
Eugene and Joan Saintz
2225 63rd Street
Fennville, MI 49408
(616) 561-2761
Fruits and vegetables in season. (C)

American Spoon Foods
411 E. Lake Street
Petoskey, MI 49770
(616) 347-9030
Pancake and waffle mix made with organically grown Indian blue corn, wild rice, wildberry preserves, wild pecans. (N)

Country Life Natural Foods
109 Oakhaven Avenue
Pullman, MI 49450
(616) 236-5011

Beans, grains, seeds, nuts, raisins. Certification varies; products labeled accordingly.

MINNESOTA
Diamond K Enterprises
R.R. 1, Box 30
St. Charles, MN 55972
(507) 932-4308
Grains, flour, cereal, pancake mixes, nuts, dried fruit, sunflower oil, honey, peanut butter, alfalfa seeds. (S)

Living Farms
Box 50
Tracey, MN 56175
(800) 622-5235 - MN
(800) 533-05320 - out-of-state
Grains, beans, rice, wheat, sunflowers, sprouting seeds (alfalfa, clover, radish). (S)

Midheaven Farms Beef
Rt. 1, Box 404
Park Rapids, MN 56470
(218) 732-4866
Beef. Ships directly to consumers in the Minnesota area only. (C)

NEW YORK
Chesnok Farm
R.D. #1, Marshland Road
Apalachin, NY 13732
(607) 748-3495
Shallots, garlic. (N)

Deer Valley Farm
R.D. 1
Guilford, NY 13780
(607) 764-8556
Beef, pork, chicken, turkey, eggs, fruit, grains, herbs, juice, pasta, oil, soup, spaghetti, spreads, seasonings, baked goods, confections, nuts. Certification varies; products labeled accordingly.

Four Chimneys Farm Winery
RD #1 HAll Road
Himrod, NY 14842
(607) 243-7502
Wine, grape juice, wine vinegar, champagne. Alcohol cannot be shipped out of state. (C)

OHIO
Millstream Marketing
1310-A East Tallmadge Ave.
Akron, Ohio 44310

(216) 630-2700
Fruits, vegetables, nuts. (N)

PENNSYLVANIA
Garden Spot Distributors
Rt. 1, Box 729A
New Holland, PA 17557
(800) 292-9631 (PA)
(800) 445-5100 (Eastern U.S.)
(717) 354-4936 (Western U.S.)
Baked goods, cereal, dried fruit, nuts, seeds, grains, flour, beans, granola, tea, herbs. (S)

Genesee Natural Foods
Rt. 449
Genesee, PA 16923-9414
(814) 228-3200 or 228-3205
Beans, seeds, flour, pasta, corn chips, rice cakes, raisins, prunes, prune and apple juice, tea, nut butters (peanut, almond, hazelnut, cashew). Certification varies; products labeled accordingly

Neshaminy Valley Natural Foods
421 Pike Road
Huntingdon Valley, PA 19006
(215) 364-8440
Grains, popcorn, beans, dried fruit, pasta, flour, cereals, seeds, nuts, miso, candy, pickles, tea, some macrobiotic products. Only a buying club of 6 households or more can order products. (S)

Rising Sun Distributors
P.O. Box 627
Milesburg, PA 16853
(814) 355-9850
Beef, poultry, lamb, pork, fruits, vegetables, beans, seeds, grains. (S)

Walnut Acres
Penns Creek, PA 17862
(717) 837-0601
Meat, fish, poultry, canned vegetables, cheese, grains, seeds, flour, nuts, pasta, seasonings, dried fruit, juice, salad dressing, granola, peanut butter. (S)

VERMONT
Hill and Dale Farms
West Hill - Daniel Davis Rd.
Putney, VT 05346
(802) 387-5817
Apples, apple vinegar. Certification pending.

VIRGINIA
Golden Acres Orchard
A.P. Thomson
Rt. 2, Box 2450
Front Royal, VA 22630
(703) 636-9611
Apples in season, apple cider vinegar, apple juice. (S)

Jordan River Farms
Huntly, VA 22640

(703) 636-9388
Grass-fed beef, free-range eggs, veal occasionally. (S)

WASHINGTON
Cascadian Farm
Star Route
Rockport, WA 98283
(206) 853-8175
Fruit conserves, dill pickles. (C)

Homestead Organic Produce
Bill Weiss
Rt. 1, 2002 Road 7 N.W.
Quincy, WA 98849
(509) 787-2248
Gourmet sweet onions. A 10 lb. minimum order is required. (S)

WEST VIRGINIA
Brier Run Farm
Rt. 1, Box 73
Birch River, WV 26610
(304) 649-2975
Soft goat cheese. (N)

Hardscrabble Enterprises, Inc.
Paul and Nan Goland
Route 6 Box 42
Cherry Grove, WV 26804
(304) 567-2727
Oak (shiitake) mushrooms. (N)

CANADA
Oak Manor Farms
Tavistock, Ontario
Canada NOB 2RO
(519) 662-2385
Grains, beans, flour, cereal, seeds, coffee. (C)

Organically Speaking
Western Europe Is Top Seed
Farmline, April 1984, pp. 16 & 17

Market days are still a tradition in Western Europe. Vendors bring fresh fruit, vegetables, and meats from nearby farms to designated city squares where shoppers often find several stands selling organically grown food.

USDA defines organic farming as food production largely without synthetically compounded fertilizers, pesticides, growth regulators, and livestock and poultry feed additives.

Organic farming has a long history in Europe and has been increasing in the last 15 years because of the characteristics of European agriculture, consumer concerns about healthy diets, product promotion, and the existence of "schools" of organic farming based on different philosophical concepts and technical practices. Although strictly organic techniques were applied to less than one percent of farmland in the ten countries of the European Community (EC) in 1982, the impact on food retailing and consumer attitudes is widespread.

One factor tending to favor organic farming in Europe is the smaller average farm size, compared with farms in the United States. A 60-to-80 acre farm is typical in Europe, while U.S. farms average more than 400 acres. Small farms can more easily produce a mixture of crops and livestock, a practice that lends itself best to organic farming because of the availability of fertilizers from animal manures and organically grown feeds.

Farms in Europe have remained small for both historical and policy reasons. Government protection of agriculture in Germany, for example, dates from the 1880's and has allowed small farms to remain viable. Laws limiting the size of livestock farms exist in Switzerland and Finland. Sweden has a number of policy measures that "protect the family farm," including limiting the number of acres per farm. Denmark and France have similar limitations, although these are rarely applied in France.

The rising cost of farm chemicals, influenced by increased energy prices, has also induced some farmers to turn to organic techniques. In many cases, yields are comparable to those achieved with nonorganic farming methods. . . .

U.S. Organics:

An estimated 25,000 U.S. farms, one percent of those with agricultural sales of $1,000 or more, produce exclusively with organic methods, while many more use some organic techniques.

The farm structure in the United States presents difficulties which, for the most part, do not exist in Europe. The last 20 years, for example, have seen an increase in average farm size here and a switch from mixed crop-livestock operations to specialized ones which lack the fertilizer and natural feed to farm or raise livestock organically. Large farm size also may inhibit the use of labor-intensive organic techniques.

The economic advantages of the smaller, mixed crop and livestock farms were enhanced as energy prices increased in the 1970's. These farms can more readily substitute organic techniques for chemicals. Some larger units have also adopted mixed crop and livestock farming and organic methods.

Marketing of organic foods appears to be more advanced in Europe than in the United States, although strictly comparable data are difficult to obtain. Many U.S. organic products are sold without special labeling or marketing and relatively few specialty shops exist. The West Coast shows the greatest interest in organic products in the United States.

Organic products face some of the same problems of legal definition in Europe as they do here. Only Oregon, Maine, and California have standards for organic labeling. Efforts at developing a nationwide certification standard for the United States were abandoned in 1973.

In the United States, 22 regional organizations exchange information among members, and certify, inspect, market, and distribute organically produced crops. The philosophical aspects of organic farming are stressed less in these groups than in Europe.

In Europe, there is both private and government involvement in certification and labeling. In France, producers, processors, and sales outlets have recently joined in an association which has obtained the legal right to set standards for labeling products as organic and for certifying farms providing organically grown produce. Among the other EC countries, only Belgium and Germany have organic labeling regulations set by the government. However, the regulations do not extend to farm certification. To date, the issue of organic labeling has not been addressed by the EC parliament, although a bill has been proposed.

Many European and American organic farm groups belong to the International Federation of Organic Agriculture Movements. The federation was formed in 1972 and has 80 member organizations in 30 countries. It promotes the aims and principles of organic agriculture, and coordinates organic farming developments internationally.

Written by economist Stephen Sposato of the International Economics Division, Economic Research Services. Sposato lived in Europe for about 10 years and worked on an organic farm in Belgium.

NUTRITION ACTION HEALTHLETTER, JAN/FEB 1985

Beef, Naturally
The Garnetts Raise Beef
Without the 'Benefit' of Antibiotics

You've heard of contented cows, but you should see the young bulls on Stephen Garnett's Virginia farm. Maybe it's the gentle lives they lead overlooking emerald pastures in the shade of the Blue Ridge Mountains. These animals are fated to wind up as "natural beef" in restaurants and food markets, but that doesn't prevent them from having an air that seems almost blissful.

There are no antibiotics in the bulls' feed, nor are there any steroid growth stimulants. But that's just part of the story. Garnett uses what he calls "Coordinated Biofarm Systems" (CBS) – a method developed by his father, Gwynn, for raising drug-free beef that he claims is superior in health, nutrition, and taste to the standard supermarket product.

The CBS system is a complete and integrated program for raising livestock. Garnett raises uncastrated bulls, which produce their own natural supply of growth hormones, rather than steers. They are fed corn, barley, and alfalfa grown in nearby fields, plus supplements such as kelp and sea minerals. The bulls are confined, but not tightly. Their feed lots are large and allow them to walk about in the open air and sunlight. Their breeding, diet, and living conditions make these animals exceptionally healthy, and there is simply no need for routine doses of antibiotics and other drugs.

"They were born here, they were raised here, and they live in an environment where they don't have any stress," Stephen Garnett explained as he gave a reporter a tour of his 520-acre operation 75 miles from Washington, D.C. Indeed, stress management is a key part of the CBS method – "TLC for life" is the way a Garnett brochure puts it.

Garnett notes that the beef is not "organic" – for example, he uses integrated pest management techniques which occasionally call for limited application of chemical pesticides on the feed crops. But the CBS method is far removed from most commercial livestock programs which include tight confinement in feed lots, regular doses of drugs and chemicals in feed, and stressful changes in routine that may impair an animal's health.

The Garnett farm is a specialty operation and his beef commands premium prices. Consequently, his operation is not strictly comparable with those of other cattlemen. But if anyone ever claims it's impossible to make a profit on beef without chemicals, send them to Stephen Garnett.

SECTION 4
For Consumer Action-Practical Alternatives

Eureka, did you know a California supermarket chain, Raley's (with 53 stores) is, through a commercial laboratory, testing 25 weekly shipments of their fresh fruit and vegetables for pesticide residues? Why aren't the giants, such as Safeway, A&P, Kroger, and Stop and Shop doing the same? The answer can be up to you and your neighbors who shop these markets. As instruments of detection become more precise and more available (someday a "sniffer" may be in wide use to locate all kinds of contaminants in food), these supermarket chains are losing their excuses. The biggest of all–Safeway–states that it does no testing of its produce nor does it intend to. The company is relying on the government to find any violations of residue standards.

More on this quest in the next section. For now, delve into the articles which give you useful information on safe kitchen practices including summer food safety advice and on avoiding pesticides through homegrown food. There follows a grab bag of shopping and eating tips (including some light on food dating codes and other definitions) and a succinct discourse on nutrition (fat, sugar, salt, fiber). And, why not, seven of our favorite recipes.

The now famous *Nutrition Guidelines*, first prepared by the U.S. Department of Agriculture ten years ago and denounced by the meat industry, are reprinted here.

Americans For Safe Food
Letter to Mary Lamielle
JUNE 17, 1987

Mary Lamielle, Director
Environmental Health Association of New Jersey
1100 Rural Avenue
Voorhees, NJ 08043

Dear Ms. Lamielle:

You asked about the Americans for Safe Food position regarding routine exterminations and other pesticide use in grocery stores and warehouses. ASF is certainly opposed to any imprudent use of pesticides that could contaminate food. As you know, most fresh produce contains pesticide residues before it even reaches the store, so there is no sense in making things worse. We understand that a store manager cannot tolerate rodents or other pests, but there are safer methods of pest control. We intend to investigate these methods more in the future, and would appreciate your help. When we have some solid recommendations, perhaps we can send them to supermarkets.

The Organic Foods Production Association of North America is working on guidelines for warehouse management and transportation (trucks are fumigated, too) in regard to organic produce. Guidelines are also needed for the cleaning of bins and shelves. The OFPANA committee working on this issue is chaired by Laurie King, Alternative Food Market, 453 Reynolds St., Oakville, Ontario L6J3M6, Canada. Her telephone number is (416) 844-2375.

David Simser, of the New Alchemy Institute (237 Hatchville Rd., East Falmouth, Massachusetts 02536, 617-563-2655) is also directly concerned about this issue. He helped design an Integrated Pest Management system for Bread & Circus, a natural food store chain in Boston, but such systems vary according to a store's needs, facilities, and particular pest problems. Simser can be hired as a consultant, but he did give us a few tips:

• Keeping storage areas clean and blocking all points of entry go a long way toward limiting pests.

• Pheromone traps are good for getting rid of certain beetles and moths.

• Pasta and some other foods can be temporarily frozen to kill pests.

• Boric acid can be used against cockroaches.

• Infested produce can be placed in a closed drum or other container along with a piece of dry ice (simple CO2) to kill bugs.

• Another source of information for safer pest control measures is a newsletter called "The IPM Practitioner," Box 7414, Berkeley, CA 94707.

Sorry it took us so long to answer your inquiry:

we've been concentrating on other things. Thank you for your concerns and good luck with your important work in New Jersey.

Sincerely,

Ben McKelway
Associate Director

Summer Food Safety Tips

BY FRANK E. YOUNG, M.D
AND KAREN J. SKINNER, PH.D.
FDA CONSUMER, JUNE 1987

Summertime, and the living is easy – not just for us, but also for microorganisms that can grow in food and make us sick.

Many view food-borne diseases as nothing more than short-term nuisances. but for certain groups – the very young, the very old, and those with impaired immune systems – food-borne illness can be serious and, tragically, often fatal. Moreover, some food-borne diseases may lead to certain chronic health problems, such as arthritis.

These sobering facts keep FDA constantly alert to the problem of microbiological food safety. But consumers are even more important than FDA in preventing food-borne disease at home, where about 30 percent of all bacteria-related food poisoning outbreaks occur. In this respect, consumers and FDA have a shared responsibility.

Summer – with its warm temperatures ideal for microbial growth – is an excellent time to review food safety principles. To help, here are a few rules to keep in mind.

Rule One: *Remember the Time-Temperature Danger Zone*

Disease-causing bacteria in food like to grow in the temperature range between 40 and 140 degrees Fahrenheit. **Avoid keeping foods in this temperature danger zone! Don't eat foods that have been kept within this range for more than two hours.** In ensuring food safety, a thermometer is the most important utensil you have. Make a practice of using it to monitor internal food temperatures and food storage temperatures.

By remembering the time-temperature relationship – the "2-40-140" rule (no more than two hours between 40 F and 140 F) – you will have a much better

idea of how to handle foods in a variety of situations, especially in the summer. If, for example, you want to buy fried chicken at a carry-out for your picnic, either eat the chicken within two hours, or take it home immediately, cool it quickly in your refrigerator, then keep it chilled below 40 F as you travel to your picnic site. If you put the hot chicken in an ice chest right away, you might not cool it rapidly enough and will actually accelerate the growth of organisms by causing the chicken to sit in the temperature danger zone for some time.

At church dinners, buffets or potluck dinners, wait as long as possible to prepare the dishes before it's time to eat, and don't leave foods sitting at room temperature. Place dishes of cold foods on beds of ice, and hold hot foods above the danger zone (at temperatures greater than 140 F). Some home-style food warmers, like chafing dishes, vary in their ability to warm food throughout. When using these warmers, don't keep food out for more than two hours. Uneven warming may lead to temperature pockets in the danger zone where bacteria happily multiply.

Use shallow dishes to cool foods quickly in the refrigerator. The interior of foods in deep containers may chill very slowly, leaving hazardous warm areas. Defrost meats and other foods, not on your kitchen counter, but in your refrigerator to avoid bacterial growth at room temperature.

If a summer electrical storm interrupts power, your refrigerator probably will remain sufficiently cold for about four to six hours (depending on room temperature), and a half-filled freezer for about one day. Keeping the refrigerator or freezer closed and using block ice in the refrigerator and dry ice in the freezer can help keep contents safely cold.

In summer, temperatures in a car can reach the high

end of the danger zone. A good rule of thumb is that perishable groceries like meats and dairy products shouldn't be left in a hot car any longer than it would take ice cream to melt. Never allow more than two hours to pass between purchasing food and getting it into your home refrigerator.

Foods available at fast-food, deli and refrigerated counters are becoming popular items for summer outings. But if they've been mishandled before you buy them, they already may have been held between 40 F and 140 F. After purchase, even two hours within the danger range for these foods might be risky. Eat them immediately or keep them cooled below 40 F.

The high temperatures (165 F to 212 F) reached in boiling, baking, frying and roasting kill most types of the bacteria that cause food-borne illness. On cook-outs, to be on the safe side, cook red meat until the pink is gone, poultry until there is no red in the joints, and fish until it's flaky.

Colder temperatures slow bacterial growth. With a thermometer, check that your freezer is at zero degrees Fahrenheit or lower and your refrigerator at 40 F or lower. Crowded summer refrigerators (like those on boats and trailers) might develop warm spots in the temperature danger range.

Rule Two: *Make a Clean Break to Good Sanitation Practices*

Cleanliness is critical to avoiding food contamination. Take extra care to avoid infecting one food with organisms from another, especially when handling raw meats and poultry, on which some bacteria usually are present. At every step of food preparation, wash hands, counters and utensils with warm, soapy water. When barbecuing, don't use the same plate for cooked meat that carried raw meat, unless you've cleaned it first. Remember, an unwashed chopping board or knife may be a reservoir of harmful bacteria. Even that favorite picnic food, watermelon, can be contaminated with a dirty utensil.

Away from home, moist hand wipes can help keep hands clean, and paper towels are one solution to the problem of dirty cloth towels harboring bacteria. If you have a cut or an infection, don't handle food. Animals aren't allowed in food processing plants, and at home, dogs, cats and other pets shouldn't be around food.

Sparkling lakes, streams and rivers tempting to thirsty campers may contain viruses, bacteria and the parasite Giardi lamblia, famous for causing "backpacker's disease" or "beaver fever." When "roughing it," boil water, treat it with purification tablets, or try one of the new filtering devices that remove Giardia cysts, as well as other contaminants. Bottled water is another alternative for drinking, cooking and cleaning.

Rule Three: *Know the Foods Requiring Special Care*

Harmful organisms grow more readily in foods high in protein and moisture. Special care with time, temperature and sanitation is needed for foods like meat, poultry, fish, shellfish, meat and seafood salads, potato salad, milk, milk products, eggs, cream pies, custards, eclairs, cream puffs, cake fillings, and gravies. Cooked pasta also can support microbial growth and should be served hot or properly refrigerated until used.

Cooking can destroy natural barriers to contamination in some foods from plant sources, and can free up nutrients needed by microorganisms to grow in these foods. Outbreaks of food-borne illness have been associated with bean curd (tofu), corn, lima beans, mushrooms, refried beans, rice, squash, and sweet potatoes that were cooked and then held for some time before eating. Except for sealed, commercially processed foods (such as canned foods), these and other moist, low-acid, cooked foods from plant sources should be refrigerated. Don't leave them at room temperature for more than two hours.

When you buy side orders of take-out food, or think about piling beans, pasta salads, or other cooked vegetables and cheeses on your salad-bar creations, remember, these are "special care foods," and consider how much time will pass before you eat.

Guarding against deadly botulism toxin, produced in reduced-oxygen environments, always requires special care. Fresh mushrooms in airtight packages recently have been involved in botulism cases. Improperly canned foods – especially low-acid foods such as meat, poultry, fish, string beans, beets, peas, corn and some fruits – also may be good places for the botulinum bacteria to grow. **Do not taste or eat** any foods from leaking, bulging or severely damaged cans; cracked jars; jars with loose or bulging lids; or swollen or puffy pouch containers. Boil all home-canned foods before tasting or serving. If an initial, rapid boil produces off-odors or foaming, **don't eat the food!** Discard suspect foods carefully, so that others, especially children and animals, won't be exposed. If there's no danger-signaling odor, boil high-acid foods for another 10 minutes and low-acid foods for another 20 minutes to destroy any botulinum toxins that nevertheless may be present.

Rule Four: *Inspect Food Storage*

Proper storage is another vital aspect of preserving the safety and quality of foods. As you restock larders at home and in summer places, take an inventory of existing items and inspect your storage practices.

Generally, pantry storage areas should be about 50 degrees Fahrenheit, clean, and away from leaky pipes, household chemicals, and openings where insects and

A Sampler of Food-Borne Illnesses

Disease and Organism That Causes It	Source of Illness	Symptoms
Salmonellosis Salmonella bacteria (more than 2,000 kinds)	Raw meats, poultry, milk and other dairy products, shrimp, frog legs, yeast, coconut, pasta and chocolate are most frequently involved.	Onset: Generally 6-48 hours after eating. Nausea, vomiting, abdominal cramps, diarrhea, fever, and headache. All age groups are susceptible, but symptoms are most severe for the elderly, infants, and the infirm.
Staphylococcal food poisoning Staphylococcal enterotoxin (produced by Staphylococcus aureus bacteria)	The toxin is produced when contaminated with the bacteria is left too long at room temperature. Meats, poultry, egg products, tuna, potato and macaroni salads, and cream-filled pastries are good environments for these bacteria to produce toxin.	Onset: Generally 1/2 to 8 hours after eating. Diarrhea, vomiting, nausea, abdominal cramps, and prostration. Mimics flu. Lasts 24-48 hours. Rarely fatal.
Botulism Botulinum toxin (produced by Clostridium botulinum bacteria)	Spores of these bacteria are widespread. However, these bacteria produce toxin only in an anaerobic (no oxygen) environment of little acidity. Botulinum toxin has been found in a considerable variety of canned foods, such as corn, peppers, green beans, soups, beets, asparagus, mushrooms, ripe olives, spinach, tuna fish, chicken, chicken liver and liver pate. It also has been found in luncheon meats, ham, sausage, stuffed eggplant, lobster, and smoked and salted fish.	Onset: Generally 4-36 hours after eating. Neurotoxic symptoms, including double vision, inability to swallow, speech difficulty, and progressive paralysis of the respiratory system. **Obtain Medical Help Immediately. Botulism Can Be Fatal!**
Campylobacteriosis Campylobacter jejuni (rod-shaped bacteria)	Bacteria found on poultry, cattle, and sheep and can contaminate the meat and milk of these animals. Chief food sources: raw poultry and meat and unpasteurized milk.	Onset: Generally 2-5 days after eating. Diarrhea, abdominal cramping, fever, and sometimes bloody stools. Lasts 7-10 days.
Perfringens food poisoning Clostridium perfringens (rod-shaped bacteria)	In most instances, the actual cause of poisoning by Clostridium perfringens is temperature abuse of prepared foods. Small numbers of the organisms are often present after cooking and multiply to food poisoning levels during cool down and storage of prepared foods. Meats and meat products are the foods most frequently implicated. These organisms grow better than other bacteria between 120 degrees - 130 degrees F; for this reason, gravies and stuffings must be kept hot, above 140 degrees F.	Onset: Generally 8-12 hours (usually 12) after eating. Abdominal pain and diarrhea. Sometimes nausea and vomiting. Symptoms last a day or less and are usually mild. Can be more serious in older or debilitated people.
Giardiasis Giardia lambia (flagellated protozoa)	Protozoa exist in the intestinal tract of humans, and cysts are expelled in feces. Giardiasis is most frequently associated with the consumption of contaminated water. It may, however, be transmitted by uncooked foods that become contaminated while growing or after cooking by infected food handlers. Cool, moist conditions favor survival of the organism.	Diarrhea (but occasionally constipation), abdominal pain, flatulence, abdominal distention, digestive disturbances, anorexia, nausea and vomiting.

rodents may enter. It's not safe to assume that all boxed and canned goods may be held at room temperature. During your inspection, check labels to ensure you've properly followed storage instructions, and discard items for which you've made a mistake. Also examine "best if used by" and expiration dates to determine if you've held foods too long. Make sure containers are free from dust and other matter that could contaminate food when products are opened.

When checking pantries in mountain cottages and other retreats, remember that cans left over the winter may freeze stressing seams and creating microscopic openings through which bacteria and other contaminants may enter. Undamaged low-acid canned goods generally last two to five years, and high-acid foods (such as tomato products and fruit juices) about 18 months.

Don't forget the refrigerator during your inspection tour. Molds, which may cause allergies and other health problems, like to grow in warm weather, but also are very content living inside refrigerators. To reduce mold buildup, wash the inside of the refrigerator with one tablespoon of baking soda dissolved in a quart of water, then rinse with clear water. Also be sure to clean the gaskets sealing the doors. Scrubbing with a solution of three tablespoons of bleach in a quart of water has been recommended for this purpose, but manufactures vary in their cleaning instructions, so consult your appliance use and care guide for recommended cleaning procedures.

Rule Five: *Think Before You Eat*

Because most food poisoning bacteria are odorless, colorless and tasteless, the only sense protecting you against food-borne illness is common sense. When traveling, prudent dietary and hygienic practices are your best safeguard against trouble. Remember these rules wherever you are, whether boating, picnicking, camping, or enjoying some other excursion.

Rule Six: *Know When to See a Doctor*

When food-borne disease strikes, see a doctor or get hospital help if the symptoms are severe or if the victim is young, elderly, or suffers from a chronic illness. If you suspect botulism, get medical help immediately! This disease can be fatal. Botulinum toxin attacks the nervous system, causing double vision, trouble swallowing, and difficulty breathing.

Generally, diarrhea, nausea, vomiting and abdominal cramps characterize food-borne illness, but symptoms vary from microbe to microbe and with the amount of contaminants actually eaten. Symptoms usually appear in six to 48 hours, but they can show up much sooner, sometimes even within half an hour. For mild cases of food poisoning, maintain liquid intake to replace fluids lost through vomiting and diarrhea.

Rule Seven: *Learn More About Food Safety*

These rules are only the highlights of food safety principles. For more information, consult these excellent sources:

• FDA's Consumer Affairs Office – Write to "Food Safety," HFE-88, Food and Drug Administration, 5600 Fishers Lane, Rockville, MD 20857.

• "The Food Keeper" – This brochure describes refrigerator and freezer storage, pantry and dry storage, and foods that need special care. For a copy, send 25 cents and a legal-sized, stamped, self-addressed envelope to: "The Food Keeper," Food Marketing Institute, 1750 K St., N.W., Washington, D.C. 20006.

• The U.S. Department of Agriculture's Meat and Poultry Hotline – Call the toll-free number, 800-535-4555, between 10 and 4 on weekdays for answers to your questions on the proper handling of meat and poultry. You may also write to "The Meat and Poultry Hotline," USDA-FSIS, Room 1165-S, Washington, D.C. 20250. Two very useful booklets, "The Safe Food Book" and "Safe Food to Go," can be obtained through the hotline.

Dr. Young is commissioner of FDA. Dr. Skinner is on temporary assignment as special assistant for science to the commissioner.

Citizen's Guide to Pesticides

SEPTEMBER 1987
United States Environmental Protection Agency
Office of Pesticides and Toxic Substances.

Home-grown Food

Growing some of your own food can be both a pleasurable activity and a way to reduce your exposure to pesticide residues in food. But, even here, there are some things you may want to do to assure that exposure is limited.

• Before converting land in an urban or suburban area to gardening, find out how the land was used previously. Choose a site that had limited (or no) chemical applications and where drift or runoff from your neighbor's activities will not result in unintended pesticide residues on your produce. Choose a garden site strategically to avoid these potential routes of entry, if possible.

If you are taking over an existing garden plot, be aware that the soil may contain pesticide residues from previous gardening activities. These residues may remain in the soil for several years, depending on the persistence of the pesticides that were used. Rather than waiting for the residues to decline naturally over time, you may speed the process.

• Plant an interim, non-food crop like annual rye grass, clover or alfalfa. Such crops, with their dense, fibrous root systems, will take up some of the lingering pesticide residues. Then discard the crops – don't work them back into the soil – and continue to alternate food crops with cover crops in the off season.

• During sunny periods, turn over the soil as often as every two to three days for a week or two. The sunlight will break down, or photodegrade, some of the pesticide residues.

Once you do begin gardening, develop strategies that will reduce your need for pesticides while maintaining good crop yields.

• Concentrate on building your garden's soil, since healthy soil grows healthy plants. Feed the soil with compost, manure, etc., to increase its capacity to support strong crops.

• Select seeds and seedlings from hardy, disease-resistant varieties. The resulting plants are less likely to need pesticides in order to flourish.

• Avoid monoculture gardening techniques. Instead alternate rows of different kinds of plants to prevent significant pest problems from developing.

• Rotate your crops yearly to reduce plant susceptibility to over-wintered pests.

• Become familiar with integrated pest management (IPM) techniques, so that you can manage any pest outbreaks that do occur without relying solely on pesticides.

• Mulch your garden with leaves, hay, grass clippings, shredded/chipped bark, or seaweed. Avoid using newspapers to keep down weeds, and sewage sludge to fertilize plants. Newsprint may contain heavy metals; sludge may contain heavy metals and pesticides, both of which can leach into your soil.

Food from the Wild

While it might seem that hunting your own game, catching your own fish, or gathering wild plant foods would reduce your overall exposure to pesticides, this isn't necessarily so. Wild foods hunted, caught, or gathered in areas where pesticides are frequently used outdoors may contain pesticide residues. Migratory species also may bear residual pesticides if these chemicals are used anywhere in their flyways.

Tolerances generally are not established or enforced for pesticides found in wild game, fowl, fish, or plants. Thus, if you consume food from the wild, you may want to take the following steps to reduce your exposure to pesticide residues.

• Although wild game is very lean and thus carries a relatively small body burden of pesticides, avoid hunting in areas where pesticide usage is very high.

• Avoid fishing in water bodies where water contamination is known to have occurred. Pay attention to posted signs warning of contamination.

• You may want to consult with fish and game officials where you plan to hunt or fish to determine whether there are any pesticide problems associated with that area.

• When picking wild plant foods, avoid gathering right next to a road, utility right-of-way, or hedgerow between farm fields which probably have been treated (directly or indirectly) with pesticides. Instead, seek out fallow fields, deep woods, or other areas where pesticide use is unlikely.

• When preparing wild foods, trim fat from meat, and discard skin of fish to remove as many fat-soluble pesticide residues as possible. For wild plant foods, follow the tips provided for commercial food. ◆

Excerpted from the Citizen's Guide to Pesticides.

Buyer's Market, Food

BY DR. MICHAEL F. JACOBSON,
VOL. 1, NUMBER 2, FEBRUARY 1985, PP. 3-7.

Nutrition:

Fat and Cholesterol

The number one nutritional problem for most Americans is the high fat and cholesterol content of their diet. Currently, 40% of the calories in the average diet come from fat. Diets high in any kind of fat increase the risk of breast, colon, and prostate cancers. Diets high in saturated fat (animal fat, butter fat, coconut and palm oils) are especially conducive to heart disease, the number one cause of death in the U.S.

Cholesterol is a fatty substance that also increases the risk of heart disease. We should eat less of it. We get about one-third of our cholesterol from eggs, one-third from dairy products (whole milk, ice cream, butter, etc.), and one-third from meat.

If you want to make one change in your diet, eat less fat and cholesterol!

Sugar

Sugar comprises about one-eighth of the calories in the average American diet. Sugar promotes tooth decay, a real pain in the mouth – and expense – for millions of people.

If you eat sugar in addition to your regular diet, the sugar will promote obesity. On the other hand, if sugar replaces other foods in your diet, you will be losing out on important nutrients.

The primary source of sugar in our diet is soda pop. Sugar is also found in an incredible variety of other foods, ranging from the obvious, like cookies and cakes, to catsup and soup.

Look for sugar on the label under such names as sucrose, dextrose, fructose, corn syrup, high-fructose corn syrup, and that old health food favorite, honey (which is really just sugar). Except for breakfast cereals, food labels generally do not disclose the

A Grab Bag of Shopping & Eating Tips

- Use unit pricing.
- Beware of the Universal Product Code (UPC) – if products are not individually priced, it makes it harder for you to compare the prices of the goods to the last time you purchased them. One way to compare is to save your receipts – with UPC, each item bought is listed by product and brand name.
- Make a list and stick to it, except to take advantage of specials.
- Clip and use coupons, but only for items that you would normally buy – you are not saving any money if you are buying items that you don't normally eat or use.
- Keep healthful snacks at the office. Avoid vending machines.
- Store bread and other perishables in refrigerator to retard spoilage.
- Consider buying day-old baked goods from local bakery outlets.
- Watch for sales; shop several local stores for good buys.
- Save re-usable plastic containers, glass jars, and paper and plastic bags.
- Be wary of food brands advertised on TV; almost all are heavily processed and overpriced.
- Try serving smaller portions of meat, the biggest part of the average food bill.
- Shop comparatively. The larger size is not always the cheapest, so be sure to check.
- Try serving meals that use whole grains and vegetables as the main course on a regular basis.
- Check prices carefully on end-of-aisle displays; what look like bargains often are not.
- To save time and energy, cook larger quantities of foods you consume regularly. Soups, casseroles, and rice keep well in the refrigerator.
- Scrutinize carefully items labeled "natural" and products parading as health food. They are often overpriced and sometimes don't provide the benefits they hint at.
- Always choose foods with the most distant expiration dates, especially on dairy products.
- Don't shop on an empty stomach, or those impulse items will find their way into your grocery cart.

Recipes

Bean Tostadas

4 onions, chopped
2 green peppers, chopped
4 cloves garlic, crushed
1-2 tbsp. olive oil
1 16 oz. can tomatoes, no salt added
1 8 oz. can tomato sauce, no salt added
3 16 oz. cans or 5 cups cooked pinto beans
1/4 tsp. cayenne (ground red pepper)
1 tbsp. oregano
1 tbsp. basil
Juice from 1/2 lemon
1/2 tsp. cumin
12 corn tortillas
Shredded cheese or chopped fresh vegetables for garnish (optional)

Saute chopped onions, green peppers, and garlic in oil in a large skillet until onions soften. Add tomatoes, tomato sauce, rinsed pinto beans, spices and lemon juice. Toast corn tortillas in oven or toaster oven until crispy (approx. 5 to 10 minutes). Serve beans on tortilla either alone or topped with shredded cheese or chopped fresh vegetables. Makes 12 tostadas.

Banana Bread

(Dry)
2 cups whole wheat flour
1/4 cup sunflower seeds, roasted
1 tbsp. baking powder, low sodium if possible
1/3 cup instant non-fat dry milk
(Wet)
3 ripe bananas, mashed
2 eggs, beaten
1/2 cup honey
2 tbsp. oil
1/3 cup water
1 tsp. vanilla

Mix dry ingredients, then wet ingredients. Combine. Pour into a greased 9" by 5" by 3" loaf pan. Bake at 350 degrees for one hour or until firm to touch.

Pasta E Fagioli (Pasta and Beans)

2 16 oz. cans or 3 cups cooked navy, Great Northern, or cannelloni (white kidney) beans
2-3 tbsp. olive oil
5 cloves garlic2 onions, chopped
3 carrots, chopped
2 stalks celery, sliced
1 32 oz. can or 8 fresh tomatoes
1 1/2 tsp. basil
1 tsp. oregano
Freshly ground black pepper to taste
1/2 lb. pasta shells
Grated parmesan cheese for garnish

Simmer cooked or canned beans in 2 cups of water with 1 tbsp. olive oil and 2 whole cloves of garlic. In a large skillet, saute onions and 3 cloves of crushed garlic in remaining olive oil. Add carrots, celery, tomatoes, and spices, cover, and cook over low heat for about 15 minutes. Salt to taste if beans are not canned in salt. Cook the paste in boiling water until it tastes slightly tender (about 10 minutes). Do not overcook. In a large pot, combine the drained pasta, sauteed vegetables, and beans with about half the bean liquid. Simmer a few minutes and serve alone or sprinkled with grated parmesan cheese. Serves 6.

Hearty Lentil Soup

2 cups dried lentils
2 medium onions, sliced
2 carrots, chopped
3-4 stalks celery, with tops, chopped
1-2 small potatoes, diced
2-3 sprigs parsley
2-3 cloves garlic, minced
1 bay leaf
2 tbsp. fresh basil or 1 tsp. dried basil
1/2 to 1 tsp. cayenne pepper
10 cups vegetable stock

Combine all ingredients in a large pot. Bring to a boil, reduce heat, and simmer for 45 minutes to one hour, until lentils are tender. Add freshly ground black pepper to taste. Serves 6-8.

Tabouli

1 1/2 cups very fine cracked wheat or bulgur
3 tomatoes, diced
1 bunch scallions (6), minced
1 large bunch parsley (1/2 cup), minced
1/4 tsp. pepper
2 tbsp. chopped mint or 1/2 tsp. dried mint (optional)
3-4 tbsp. olive oil
Juice of 2 lemons (or more, to taste)

Soak cracked wheat for 1/2 hour to 1 hour (may need 3-4 hours if not very fine wheat) in 4 cups of water, then remove excess water. Toss thoroughly with all ingredients, taste for seasoning. Chill. Tastes great with lettuce or in pita (pouch) bread. Serves 6.

Muffins*

(Dry)
2 cups whole wheat flour
2 tsp. low-sodium baking powder
(Wet)
1 egg, beaten
2 tbsp. vegetable oil
1/4 cup honey
1 1/2 cups buttermilk

Preheat oven to 400 degrees. Mix the dry ingredients together. Mix the wet ingredients together. Combine. Spoon into 12 large greased muffin cups. Bake for 15 to 20 minutes. Makes 12 muffins.

*Most people think buttermilk is as rich as butter. In fact, buttermilk has no more fat or calories than skim milk, from which it is usually made. Unfortunately, salt is added to most brands of buttermilk. Otherwise, it is an ideal ingredient for baked goods.

Baked Potato or Sweet Potato

Nothing's easier! Bake in oven at 450 degrees for 50 minutes. Top with herbs, a little margarine, or milk, or yogurt, or eat it plain.

amount of sugar in a product.

Salt/Sodium

Most processed foods are laced with salt (sodium chloride) or other sodium-containing additives (e.g., monosodium glutamate, sodium benzoate). The salt flavors the foods, but, unfortunately, too much sodium increases the risk or severity of high blood pressure (hypertension).

Americans get about 65% of their sodium from processed foods, about 25% from salt shakers and home cooking, and the rest from natural foods.

One of the most important dietary changes to make is to eat less sodium. Instead of salt, you can season foods with lemon juice, herbs, and other condiments.

Look for no-salt-added processed foods – or, better yet, stick to natural foods.

Dietary Fiber

Dietary fiber, or roughage, is something the body cannot digest, but which serves important functions on its journey through your body. People who eat too little fiber – and that includes most Americans – often suffer from constipation and other disorders of the digestive tract. Diets high in fiber appear to reduce the risk of colon (large intestine) cancer and low blood cholesterol levels slightly.

All vegetable-based foods contain dietary fiber, unless it has been largely removed, as in the case of white flour and white rice. Beans (lentils, kidney beans, navy beans, etc.), whole grains, and bran are several of the best sources of fiber.

Food Additives

Once upon a time, food was food. It was stuff that grew in fields, on trees, or on ranches. Now, many foods have a heritage not just in the soil, but in the chemistry lab.

Packaged foods generally list ingredients on the label. Hundreds of chemical additives may be used to extend shelf life, prevent bacterial growth, or make a food appear fresher or more nutritious than it really is. The following list describes some of the most questionable preservatives, flavorings, and colorings. Avoiding these additives will not only help you avoid the main toxic concerns, but will also steer you away from most junk foods. To put things into perspective, though, diets high in fat, cholesterol, and sodium pose a much bigger health problem than food additives.

"Least Wanted" Food Additives

Artificial Colorings (cancer risk, allergic reactions) Citrus Red #2, Red #40, Yellow #5, Blue #2.

BHA/BHT (cancer risk) Unnecessary anti-oxidant preservatives that are added to many cereals, potato chips, and oils to extend their shelf lives.

Caffeine (breast lumps, birth defects, insomnia) A stimulant that occurs naturally in coffee and tea, but that is added to soda pop.

Monosodium Glutamate (MSG) ("Chinese Restaurant Syndrome" – headache, tingling, muscle tightness) A widely-used flavor enhancer found in many processed foods.

Propyl Gallate (cancer risk) This anti-oxidant preservative extends the shelf life of vegetable oil, processed meat, potato sticks, soup base, and chewing gums.

Quinine (poorly tested for ability to cause cancer or birth defects) A flavoring added to soft drinks.

Saccharin (cancer risk) A popular artificial sweetener.

Sodium Nitrite (cancer risk) An artificial coloring, flavoring, and preservative added to bacon, ham, hot dogs, and luncheon meats.

Sulfiting Agents (severe asthmatic reactions) Used to prevent discoloration and bacterial growth in dried apricot, beer, restaurant salads, shellfish, wine, grape juice, dehydrated potatoes.

A Primer on Food Labels

EXCERPTS FROM

The Goldbecks' Guide to Good Food

BY NIKKI AND DAVID GOLDBECK

(NAL BOOKS: NEW YORK, 1987)

Food Dating Codes

Even the notorious Jack Benny couldn't conceal his true age as effectively as many prepared foods do. Food dating is not regulated by the federal government; however, some states and localities have their own requirements for certain products (e.g., milk). Traditionally, most dates on food packages have been written in code for use by industry. Many companies do voluntarily offer information to the consumer, however.

Open Dating

When the date is provided for consumers, it can mean one of four different things.

Pack date is the day the product was manufactured, processed, or packaged. It tells how old the product is. A pack date is usually found on packaged foods with a long shelf–life.

Pull or sell date is the last day the product should be sold, assuming it has been handled properly. It allows time for home consumption, but how much time is not revealed. Cold cuts, dairy products, and bakery goods most commonly use pull or sell date.

Expiration date is the last day the food should be eaten. Infant formula and yeast are examples of foods that carry expiration dates.

Freshness date or use–before date is the time in which the product remains at highest quality. Sometimes foods are sold at a reduced price for a few days after this.

Blind Dates

While there is no single format used in commercial manufacturing codes, there is almost always a hidden indication of date of processing or packaging.

The numbers can be as straightforward as 62685 representing June 26, 1985, but it is just as likely that the first number indicates the year, so that 5355 could be interpreted as the 355th day of 1985 (or December 20, 1985). Or, the first two digits could mean the week of manufacture and the last year. For example, 016 is deciphered as the first week of 1986, or January 1–7, 1986. Unfortunately, other numbers may be inserted in the code to represent the product, shift, plant, hour, etc.

If you really want to know about a particular product, check with the store manager or write to the manufacturer.

Ambiguous Language

Some of the descriptions on the label that are totally unregulated may hold quite different meanings for the manufacturer and the consumer. Until certain terms are clarified, a wise shopper should look behind these statements in order to steer clear of unscrupulous merchandisers.

Natural

A 1984 government survey found that 63 percent of those polled believed "natural" foods were more nutritious and 47 percent were willing to pay a 10 percent premium for these foods. If some uniform meaning were ascribed to the word "natural," these attitudes might be justified. As it stands, though, "natural" has not been given an official government definition, except for meat.

Indeed, "natural" cheeses contain artificial coloring and preservatives; "natural" ice creams are not only loaded with sugar but may include such untraditional ingredients as vegetable gum and mono– and diglycerides; "natural" breads are often made of refined white flour and are fortified with synthetic nutrients; some "natural" cereals are more highly sweetened than processed varieties; foods that are "naturally flavored" may also contain artificial flavoring, preservatives, thickeners, and colorants. Even margarine, a totally fabricated product, has been described as "natural" by some manufacturers.

Some "natural food" companies also include white flour, hydrogenated shortening, hydrolyzed vegetable protein, soy protein isolates, vegetable gums, and refined sweeteners in their products.

In all fairness to the consumer, a "natural product" should have all ingredients fully identified, including minor ingredients within ingredients. In addition, the condition of each ingredient should be described:

whether it has been bleached, dehydrated or dried, hydrogenated, defatted (using a solvent), rolled, ground, etc. Processing should be limited to what could be done in a typical, well–equipped home kitchen.

Natural Meat

The USDA does restrict use of the term "natural" for meat and poultry to those that "do not contain any artificial flavor or flavoring, coloring ingredient or chemical preservative, or any other artificial or synthetic ingredient, and the product and its ingredients are not more than minimally processed." If the meat or poultry product contains all natural ingredients, but they are not normal constituents of the product in question — e.g., beet powder used to color gravy — the statement "all natural ingredients" must be used in lieu of "all natural." Even this is not totally reliable, because sugar, large amounts of salt, hydrogenated fats, textured vegetable protein, and similarly highly processed ingredients do appear in some products the USDA allows the manufacturer to call "natural."

Organic

"Organic" is no less troubling than "natural." Technically, "organic" means "chemical compounds containing carbon." Many agricultural chemicals are carbon–based and could meet this definition. However, the term as popularized by J.I. Rodale is intended to mean something else.

Most authorities agree that "organically grown" food means food that has been raised without pesticides, synthetic fertilizers, or other related chemicals, in soil that is nourished only with the addition of live matter and mineral fertilizers. While the federal government places no such limits on the terms "organic," some state governments do. As of 1986, California, Oregon, Maine, Washington, Minnesota, Massachusetts, Nebraska, and Montana had established "organic" laws.

"Organically processed" food, in addition to being organically grown, should not be treated with any artificial or synthetic additives in its preparation.

When "organic" is applied to animal–derived foods it means that the feed and water were free of pesticides and chemical contaminants, that there has been no routine use of growth stimulants, hormones, or other drugs, and that the latter were used only as required for treatment of an illness and then within a restricted time from slaughter or milking. The USDA does not permit the word "organic" on meat and poultry labels, but foods meeting these specifications may be labeled "naturally grown."

Organic farm organizations, processors, and distributors have joined together in a number of trade organizations. These groups run certification programs which include annual farm visits, sworn affidavits, and detailed questionnaires as to growing, harvesting, and storage practices.

Nutrition and Your Health

DIETARY GUIDELINES FOR AMERICANS, 2ND EDITION, 1985
U.S. DEPARTMENT OF AGRICULTURE
U.S. DEPARTMENT OF HEALTH AND HUMAN SERVICES

What should you eat to stay healthy?

The life expectancy, average body size, and general good health of the American population seem to indicate that most diets are adequate. Foods we have to choose from are varied, plentiful, and wholesome.

Even so, hardly a day goes by without someone trying to tell us what we should and should not eat. Newspapers, magazines, books, radio, and television give us lots of advice. Unfortunately, much of it is confusing.

Some of this confusion exists because we don't know enough about nutrition to identify an "ideal diet" for each individual. People differ – and their food needs differ depending on age, sex body size, physical activity, and other conditions such as pregnancy and illness.

In those chronic conditions where diet may be

important – heart disease, high blood pressure, strokes, tooth decay, diabetes, osteoporosis, and some forms of cancer – the roles of specific dietary substances have not been defined fully.

Research seeks more information about the amounts of essential nutrients people need and diet's role in certain chronic diseases. Much attention has been devoted recently, for example, to the possible effects of calcium intake on osteoporosis, and of dietary fat and fiber on certain forms of cancer and heart disease.

But what about advice for today? The following guidelines tell how to choose and prepare foods for you and your family. This advice is the best we can give based on the nutrition information we have now.

The first two guidelines form the framework for a good diet: "Eat a variety of foods" that provide enough of essential nutrients and energy (calories) to "maintain desirable weight." The next five guidelines describe special characteristics of good diets. They suggest that you get adequate starch and fiber and avoid too much fat, sugar, sodium, and alcohol.

The Recommended Dietary Allowances (RDA) are suggested amounts of energy, protein, and some minerals and vitamins for an adequate diet. For other dietary substances, specific goals must await further research. However, for the U.S. population as a whole, increasing starch and fiber in our diets and reducing calories (primarily from fats, sugars, and alcohol) is sensible. These suggestions are especially appropriate for people who have other risk factors for chronic diseases, such as family history of obesity, premature heart disease, diabetes, high blood pressure, high blood cholesterol levels, or for those who use tobacco, particularly cigarette smokers.

The guidelines are suggested for most Americans – those who are already healthy. They do not apply to people who need special diets because of diseases or conditions that interfere with normal nutritional requirements. These people may need special instruction from registered dietitians, in consultation with their own physicians.

No guidelines can guarantee health and well-being. Health depends on many things, including heredity, lifestyle, personality traits, mental health and attitudes, and environment, in addition to diet.

Food alone cannot make you healthy. But good eating habits based on moderation and variety can help keep you healthy and even improve your health.

Dietary Guidelines for Americans
- Eat a variety of foods.
- Maintain desirable weight.
- Avoid too much fat, saturated fat, and cholesterol.
- Eat foods with adequate starch and fiber.
- Avoid too much sugar.
- Avoid too much sodium.
- If you drink alcoholic beverages, do so in moderation.

Eat a Variety of Foods
You need more than 40 different nutrients for good health. These include vitamins and minerals, amino acids (from proteins), essential fatty acids (from fats and oils), and sources of energy (calories from carbohydrates, fats, and proteins). Adequate amounts of these nutrients are present in the foods in a well-balanced diet.

Most foods contain more than one nutrient. For example, milk provides protein, fats, sugar, riboflavin and other B vitamins, vitamin A, calcium, phosphorus, and other nutrients; meat provides protein, several B vitamins, iron, and zinc in important amounts.

Except for human milk during the first 4 to 6 months of life, no single food item supplies all of the essential nutrients in the amounts that you need. Milk, for instance, contains very little iron and meat provides little calcium. Thus, you should eat a variety of foods to get an adequate diet. With a variety of foods, you are more likely to get all the nutrients you need.

One way to assure variety – and with it, a well-balanced diet – is to select foods each day from each of the major food groups. These groups include: fruits; vegetables; cereals and other foods made from grains, such as breads; milk and dairy products such as cheese and yogurt; and meats, fish, poultry, eggs, and dry beans and peas. Select different foods from within groups, too.

Fruits and vegetables are good sources of vitamin A, vitamin C, folic acid, fiber, and many minerals. Whole-grain and enriched breads, cereals, and other grain products provide B vitamins, iron, protein, calories, and fiber. Meats, poultry, fish and eggs supply protein, fat, iron, and other minerals, as well as several B vitamins. Dairy products are major sources of calcium and many other nutrients. Recent research suggests that calcium may play a role in preventing osteoporosis.

To Assure Yourself an Adequate Diet
Eat a variety of foods daily in adequate amounts, including selections of:
- Fruits;
- Vegetables;
- Whole-grain and enriched breads, cereals, and other products made from grains;
- Milk, cheese, yogurt and other products made from milk; and
- Meats, poultry, fish, eggs, and dry beans and peas.

The number and size of portions should be adjusted to reach and maintain your desirable body weight.

There are no known advantages and some potential harm in consuming excessive amounts of any nutrient. Large dose supplements of any nutrient should be avoided.

You will rarely need to take vitamin or mineral supplements if you eat a variety of foods. There are a few important exceptions to this general statement:

- **Women in their childbearing years** may need to take iron supplements to replace the iron they lose with menstrual bleeding. Women who are no longer menstruating should not take iron supplements routinely.

- **Women who are pregnant or who are breast-feeding** need more of many nutrients, especially iron, folic acid, vitamin A, calcium, and sources of energy. Detailed advice should come from their physicians and dietitians.

- **Infants** also have special nutritional needs. Infants should be breast-fed unless there are special problems. The nutrients in human breast milk tend to be absorbed by the body better than those in cow milk or infant formula. In addition, breast milk serves to transfer immunity to some diseases from the mother to the infant.

Normally, most babies are not given solid foods until they are 3 to 6 months old. At that time, solid foods can be introduced gradually. Prolonged breast- or formula-feeding – without solid foods or supplemental iron – may result in iron deficiency.

Salt or sugar should not be added to the baby's foods. Extra flavoring with salt and sugar is not necessary – infants do not need these inducements if they are really hungry.

To Assure Your Baby an Adequate Diet
- Breast-feed unless there are special problems.
- Delay other foods until baby is 4 to 6 months old.
- Do not add salt or sugar to baby's food.

- **Elderly people** may eat relatively little food. Thus, they need to eat less of foods that are high in calories and low in essential nutrients – such as fats and oils, sugars and sweets and alcohol. (Alcohol often is not thought of as a food, but is high in calories.)

Elderly people who eat a varied diet do not generally need vitamin and mineral supplements. However, some medications used for the treatment of diseases may interact with nutrients. In such instances, a physician may prescribe supplements.

Maintain Desirable Weight

If you are too fat, your chances of developing some chronic disorders are increased. Obesity is associated with high blood pressure, increased levels of blood fats (triglycerides) and cholesterol, heart disease, strokes, the most common types of diabetes, certain cancers, and many other types of ill health. Thus, you should try to maintain a "desirable" weight.

But, how do you determine what a desirable weight is for you?

There is no absolute answer. The table shows desirable ranges for most adults. If you have been obese since childhood or adolescence, you may find it difficult to reach or maintain your weight within a desirable range. Generally, the weight of adults should not be much more than it was when they were younger – about 25 years old.

It is not well understood why some people can eat much more than others and still maintain desirable weight. However, one thing is definite – to lose weight, you must take in fewer calories than you burn. This means that you must either choose foods with fewer calories or you must increase your physical activity, preferably both.

To Help Control Overeating
- Eat slowly
- Take smaller portions
- Avoid "seconds"

For most people who decide to lose weight, a steady loss of 1 to 2 pounds a week – until you reach your goal – is safe. At the beginning of a weight-reduction diet, much of your weight loss comes from loss of water. Long-term success depends on new and better habits of eating and exercise. That is why so-called "crash" and "fad" diets usually fail in the long run.

Do not try to lose weight too rapidly. Avoid crash diets that are severely restricted in the variety of foods they allow. Diets containing fewer than 800 calories may be hazardous and should be followed only under medical supervision. Some people have developed kidney stones, disturbing psychological changes, and other complications while following such diets. A few people have died suddenly and without warning.

Also, do not attempt to lose weight by inducing

Desirable Body Weight Ranges

Height Without Shoes	Weight Without Clothes	
	MEN	*WOMEN*
(FEET-INCHES)	(POUNDS)	(POUNDS)
4'10"		92-121
4'11"		95-124
5'0"		98-127
5'1"	105-134	101-130
5'2"	108-137	104-134
5'3"	111-141	107-138
5'4"	114-145	110-142
5'5"	117-149	114-146
5'6"	121-154	118-150
5'7"	125-159	122-154
5'8"	129-163	126-159
5'9"	133-167	130-164
5'10"	137-172	134-169
5'11"	141-177	
6'0"	145-182	
6'1"	149-187	
6'2"	153-192	
6'3"	157-197	

Note: For women 18-25 years, subtract one pound for each year under 25.
Source: Adapted from the 1959 Metropolitan Desirable Weight Table

vomiting or by using laxatives. Frequent vomiting and purging can cause chemical imbalances which can lead to irregular heartbeats and even death. Frequent vomiting can also erode tooth enamel. Avoid these and other extreme means of losing weight.

To Lose Weight
• Eat a variety of foods that are low in calories and high in nutrients:
• Eat more fruits, vegetables, and whole grains
• Eat less fat and fatty foods
• Eat less sugar and sweets
• Drink less alcoholic beverages
• Increase your physical activity

A gradual increase of everyday physical activity like brisk walking can also be very helpful in losing weight and keeping it off. The [accompanying] chart gives the approximate calories used per hour in different activities.

Do not attempt to reduce your weight below the desirable range. Severe weight loss may be associated with nutrient deficiencies, menstrual irregularities, infertility, hair loss, skin changes, cold intolerance, severe constipation, psychiatric disturbances, and other complications.

If you lose weight suddenly or for unknown reasons, see a physician. Unexplained weight loss may be an early clue to an unsuspected underlying disorder.

Avoid Too Much Fat, Saturated Fat, and Cholesterol
If you have a high blood cholesterol level, you have a greater chance of having a heart attack. Other factors can also increase you risk of heart attack – high blood pressure and cigarette smoking, for example – but high blood cholesterol is clearly one of the major risk factors.

Populations like ours with diets relatively high in fat (especially saturated fat) and cholesterol tend to have high blood cholesterol levels. Individuals within these populations have a greater risk of having heart attacks than individuals within populations that have diets containing less fat.

Eating extra saturated fat, high levels of cholesterol, and excess calories will increase blood cholesterol in many people. Of these, saturated fat has the greatest influence. There are, however, wide variations among individuals– related to heredity and to the way each person's body uses cholesterol.

Some people can have diets high in saturated fats and cholesterol and still maintain desirable blood cholesterol levels. Other people, unfortunately, have high blood cholesterol levels even if they eat low-fat, low-cholesterol diets. However, as noted above, for many people, eating extra saturated fat, high levels of cholesterol, and excess calories will increase blood cholesterol.

There is controversy about what recommendations are appropriate for healthy Americans. But for the U.S. population as a whole, it is sensible to reduce daily consumption of fat. This suggestion is especially appropriate for individuals who have other cardiovascular risk factors, such as smokers or those with family histories of premature heart disease, high blood pressure, and diabetes.

The recommendations are not meant to prohibit you from using any specific food item or to prevent you from eating a variety of foods. Many foods that contain fat and cholesterol also provide high quality protein and many essential vitamins and minerals. You can eat these foods in moderation, as long as your overall fat and cholesterol intake is not excessive.

To Avoid Too Much Fat, Saturated Fat, And Cholesterol
• Choose lean meat, fish, poultry, dry beans and peas as your protein sources
• Use skim or low-fat milk and milk products
• Moderate your use of egg yolks and organ meats (such as liver)
• Limit your intake of fats and oils, especially those high in saturated fat, such as butter, cream, lard, heavily hydrogenated fats (some margarines), and foods containing palm and coconut oils
• Trim fat off meats
• Broil, bake, or boil rather than fry
• Moderate your use of foods that contain fat, such as breaded and deep-fried foods
• Read labels carefully to determine both amount and type of fat present in foods

Eat Foods With Adequate Starch and Fiber
The major sources of energy (calories) in the American diet are carbohydrates and fats. (Protein and alcohol also supply calories.) Carbohydrates are especially helpful in weight-reduction diets because, ounce for ounce, they contain about half as many calories as fats do.

Simple carbohydrates, such as sugars, and complex carbohydrates, such as starches, have about the same caloric content. But most foods high in sugar, such as candies and other sweets, contain little or no vitamins and minerals. On the other hand, foods high in starch, such as breads and other grain products, dry beans and peas, and potatoes, contain many of these essential nutrients.

Eating more foods containing complex carbohydrates can also help to add dietary fiber to your diet. The American diet is relatively low in fiber.

Dietary fiber is a term used to describe parts of plant foods which are generally not digestible by humans.

There are several kinds of fiber with different chemical structures and biological effects. Because foods differ in the kinds of fiber they contain, it's best to include a variety of fiber-rich foods – whole-grain breads and cereals, fruits, and vegetables, for example.

Eating foods high in fiber has been found to reduce symptoms of chronic constipation, diverticular disease, and some types of "irritable bowel." It has also been suggested that diets low in fiber may increase the risk of developing colon cancer. Whether this is true is not yet known.

How dietary fiber relates to cancer is one of many fiber topics under study. Some others are the fiber content of foods and the amount of fiber we need in our diets. Also being studied are whether fiber extracted from food has the same effect as that from intact food and the extent to which high fiber intakes may lead to trace mineral deficiency.

Advice for today: A diet containing whole-grain breads and cereals, fruits, and vegetables should provide an adequate intake of dietary fiber. Increase your fiber intake by eating more of these foods that contain fiber naturally, not by adding fiber to foods that do not contain it.

To Eat More Starch and Fiber
• Choose foods that are good sources of fiber and starch, such as whole-grain breads and cereals, fruits, vegetables, and dry beans and peas.
• Substitute starchy foods for those that have large amounts of fats and sugars

Avoid Too Much Sugar
A significant health problem from eating too much sugar is tooth decay (dental caries). The risk of caries is not simply a matter of how much sugar and sugar-containing foods you eat but how often you eat them. The more frequently you eat sugar and sugar-containing foods, the greater the risk for tooth decay – especially if they are eaten between meals, and they stick to your teeth.

Americans consume sugar in various forms in their diets. Common table sugar (sucrose) is only one form of sugar. Other sugars – such as glucose (dextrose), fructose, maltose, and lactose – occur naturally in foods and are added as ingredients in foods, e.g. corn sweeteners. Both starches and sugars appear to increase the risk of tooth decay when eaten between meals, but simple sugars appear to offer a higher risk. Thus, frequent in-between-meal snacks of foods such as cakes and pastries, candies, dried fruits, and soft drinks may be more harmful to your teeth than the sugars eaten in regular meals.

You cannot avoid all sugar because most of the foods we eat contain some sugar in one form or another. But keep the amount of sugars and sweet foods you eat moderate. And when you do eat them, brush your teeth afterwards, if possible.

Clearly, there is more to maintaining healthy teeth than avoiding sugars. Careful dental hygiene and exposure to adequate amounts of fluoride through fluoridated water are especially important. Fluoridated toothpastes or mouth rinses are helpful, particularly where there is no fluoridated water.

Contrary to widespread belief, too much sugar in your diet does not cause diabetes. The most common type of diabetes occurs in obese adults; avoiding sugar without correcting the overweight problem – which requires reduction in total caloric intake – will not

Approximate Energy Expenditure by a Healthy Adult Weighing About 150 Pounds

Activity	Calories per hour
Lying quietly	80-100
Sitting quietly	85-105
Standing quietly	100-120
Walking slowly 2-1/2 mph	210-230
Walking quickly, 4 mph	315-345
Light work, such as ballroom dancing; cleaning house; office work; shopping	125-310
Moderate work, such as cycling, 9 mph; jogging, 6 mph; tennis; scrubbing floors; weeding garden	315-480
Hard work, such as aerobic dancing; basketball; chopping wood; cross-country skiing; running, 7 mph; shoveling snow; spading garden; swimming, "crawl"	480-625

Source: Based on material prepared by Robert E. Johnson, M.D., Ph.D., and colleagues, Professor Emeritus, University of Illinois.

solve the problem.

Sugars provide calories but few other nutrients. Thus, diets with large amounts of sugars should be avoided, especially by people with low calorie needs, such as those on weight-reducing diets and the elderly.

To Avoid Too Much Sugar

• Use less of all sugars and foods containing large amounts of sugars, including white sugar, brown sugar, raw sugar, honey, and syrups. Examples include soft drinks, candies, cakes, and cookies

• Remember, how often you eat sugar and sugar-containing food is as important to the health of your teeth as how much sugar you eat. It will help to avoid eating sweets between meals

• Read food labels for clues on sugar content. If the name sugar, sucrose, glucose, maltose, dextrose, lactose, fructose, or syrups appears first, than there is a large amount of sugar

• Select fresh fruits or fruits processed without syrup or with light, rather than heavy, syrup

Avoid Too Much Sodium

Table salt contains sodium and chloride – both are essential in the diet. In addition, salt is often required for the preservation of certain foods.

Sodium is present in many beverages and foods that we eat, especially in certain processed foods, condiments, sauces, pickled foods, salty snacks, and sandwich meats. Baking soda, baking powder, monosodium glutamate (MSG), and even many medications (many antacids, for instance) contain sodium.

A major hazard for excess sodium is for persons who have high blood pressure. Not everyone is equally susceptible. In the United States, about one in four adults has elevated blood pressure. Sodium intake is but one of the factors known to affect high blood pressure. Several other nutrients may also be involved. Obesity plays a major role.

In populations with low sodium intakes, high blood pressure is rare. In contrast, in populations with high sodium intakes, high blood pressure is common. If people with high blood pressure severely restrict their sodium intakes, their blood pressure will usually fall, although not always to normal levels.

At present, there is no good way to predict who will develop high blood pressure, although certain groups such as blacks have a higher prevalence. Low-sodium diets may help some people avoid high blood pressure if they could be identified before they develop the condition.

Since most Americans eat more sodium than is needed, consider reducing your sodium intake. Use less table salt, read labels carefully, and eat sparingly those foods to which large amounts of sodium have been added. Remember that a substantial amount of the sodium you eat may be "hidden" – either occurring naturally in foods or as part of a preservative or flavoring agent that has been added.

To Avoid Too Much Sodium

• Learn to enjoy the flavors of unsalted foods

• Cook without salt or with only small amounts of added salt

• Try flavoring foods with herbs, spices, and lemon juice

• Add little or no salt to food at the table

• Limit your intake of salty foods such as potato chips, pretzels, salted nuts and popcorn, condiments (soy sauce, steak sauce, garlic salt), pickled foods, cured meats, some cheeses, and some canned vegetables and soups

• Read food labels carefully to determine the amounts of sodium

• Use lower sodium products, when available, to replace those you use that have higher sodium content

If You Drink Alcoholic Beverages, Do So in Moderation

Alcoholic beverages are high in calories and low in nutrients. Thus, even moderate drinkers will need to drink less if they are overweight and wish to reduce.

Heavy drinkers frequently develop nutritional deficiencies as well as more serious diseases, such as cirrhosis of the liver and certain types of cancer, especially those who also smoke cigarettes. This is partly because of loss of appetite, poor food intake, and impaired absorption of nutrients.

Excessive consumption of alcoholic beverages by pregnant women may cause birth defects or other problems during pregnancy. The level of consumption at which risks to the unborn occur has not been established. Therefore, the National Institute on Alcohol Abuse and Alcoholism advises that pregnant women should refrain from the use of alcohol.

One or two standard-size drinks daily appear to cause no harm in normal, healthy, nonpregnant adults. Twelve ounces of regular beer, 5 ounces of wine, and 1-1/2 ounces of distilled spirits contain about equal alcohol.

If you drink, be moderate in your intake and DO NOT DRIVE!

Acknowledgements: The U.S. Department of Agriculture and the U.S. Department of Health and Human Services acknowledge the recommendations of the Dietary Guidelines Advisory Committee which were the basis of this revision. The committee consisted of Dr. Bernard Schweigert (chairman), Dr. Henry Kamin, Dr. David Kritchevsky, Dr. Robert E. Olson, Dr. Lester Salans, Dr. Robert Levy, Dr. Sanford A. Miller, Dr. Judith S. Stern, and Dr. Frederick J. Stare.

For a list of materials on how to use the Dietary Guidelines, write to Human Nutrition Information Service, U.S. Department of Agriculture, Room 325 A, Federal Building, Hyattsville, Maryland 20782.

For additional help with diet and health questions write to Consumer Inquiries, Food and Drug Administration, 5600 Fishers Lane, Rockville, Maryland 20857 or contact the dietitian, home economist, or nutritionist in the following groups:

  Public Health Department
  County Extension Service
  State or Local Medical Society
  Hospital Outpatient Clinic
  Local American Red Cross Chapter
  Local Dietetic Association
  Local Diabetes Association
  Local Heart Association
  Local Health Center or Clinic

Note: These dietary guidelines are intended only for populations with food habits similar to those of people in the United States.

Home and Garden Bulletin No. 232

SECTION 5
For Citizen Action–Practical Alternatives

This is largely the program of Americans for Safe Food – and it is indeed practical. Learn what to negotiate with your supermarket and about success stories. The excerpts here are just to perk your interest to write Americans for Safe Food about their projects and their colorful posters and pamphlets. They will send you free the entire manual for local safe food organizers. Excerpts from the Food Marketing Institute's interview with Dr. Michael Jacobson outline his strategies of action. His Center has started honoring "Safe Food Trailblazers" and the first list is reprinted in this section.

A second campaign is the farmworker-consumer boycott of table grapes in order to force "agribusiness to stop poisoning workers and consumers." The long-time leader of the United Farm Workers of America, Cesar Chavez, has written to millions of Americans to enlist their help. The letter is reprinted in this section. In 1970, Chavez and his farmworkers launched a grape boycott which resulted in the growers agreeing not to use DDT, DDE and Dieldrin and the signing of a fair labor contract with the workers who harvest the grapes. This letter by Chavez is not one you'll soon forget.

Interview with
Dr. Michael Jacobson

Executive Director
Center for Science in the Public Interest
FOOD MARKETING INSTITUTE ISSUES BULLETIN, SEPTEMBER 1986

Dr. Jacobson holds a Ph.D. in microbiology from the Massachusetts Institute of Technology. He has published technical papers in many scientific journals and articles in major newspapers and popular magazines. He also has appeared on numerous national TV and radio shows....

...CSPI is a non-profit organization in Washington, D.C., that "seeks to provide the public with reliable, interesting, understandable information about food, the food industry, and the government regulation of food," according to CSPI's publicity. Its goal is "to improve the quality of the American diet through research and public education...."

Q: What factors led to launching the Americans for Safe Food campaign at this time? Tell us your goals, activities and view of the supermarket role.
A: The campaign's goal is to educate people about the problems. We want to put them into context, focusing on the larger problems first and then increasing the availability of pure, safe food.

This will be a long, slow process. We would like to have some food that is grown without these chemicals, more food with only some of them and then clear labeling. Consumers could be told, for example, if a particular brand of poultry is grown without antibiotics.

We would like private-label or brand-name products labeled – maybe not listing every chemical but some of the priorities, such as antibiotics and Alar.

Our main activities will involve working with consumer and environmental groups and concerned citizens at the local level. We will encourage them to initiate a grass-roots movement for safe food.

We are suggesting a wide range of activities, from advocating laws concerning organically grown foods to labeling products containing chemicals and chemical residues.

We will conduct education programs and distribute leaflets. We will ask supermarkets to provide more healthful products. We will start by meeting with the supermarkets, and if they are uncooperative, we will try more public and aggressive activities. Local groups will decide on their own activities.

The supermarket is the focal point for this kind of campaign, because it is a centralized seller to consumers and purchasing agent from suppliers.

Many supermarkets have private-label products with which they can do a lot more than they can ask of their suppliers. Some supermarkets have already put their toes in the water.

The campaign's main focus is not government

regulation to provide pure food to all consumers. Its focus will be to try to put pressure on the marketplace to generate demand.

I do think suppliers will comply. In fact many farmers already grow food essentially without pesticides, hormones, and antibiotics. There is no marketing system for them, however – no way to channel their products to stores. We need those systems in place to help these growers.

State governments can help. Jim Hightower, commissioner of agriculture in Texas, has been encouraging growers' cooperatives to market to large supermarkets.

Small farmers may be able to grow pesticide-free foods more easily than larger farmers. If they can get a corner on the marketplace, they may be able to survive. As it is, many small farms are going out of business so fast they do have the chance to experiment.

Q: You mentioned labeling. What do you have in mind?

A: In some cases, the suppliers could apply labels to their products. In other cases the supermarket could put up a sign. For example, a sign could say, "These grapes are organically grown."

Q: If your premise is that contaminant-free products may cost more, what about individuals who will not be able to afford the higher costs?

A: They will need to have a choice if there is a cost difference. Hopefully the greater the supply of pure, safe foods, the greater the competition – and the smaller the cost differential.

Q: CSPI initiated the campaign to require fast-food companies to label ingredients in their foods. What is your assessment of their efforts so far?

A: The response has been limited and begrudging. There has been no labeling of fast foods.

A few of the larger companies have been persuaded by state attorneys-general to provide brochures listing ingredients for their foods. That is a major breakthrough. Until this ...[1986] almost all the companies refused to disclose the ingredients. They said the ingredients were trade secrets.

About six companies have said they will provide pamphlets in their stores. Considering they sell to millions of people a day, enough pamphlets could not be supplied to meet the need. It is an inefficient way to provide information to the public.

We are still pressing for ingredient labeling on the fast-food packages. That would be inexpensive and quite feasible. There is legislation in both the House and the Senate to require this. We are hoping for a hearing this fall – and passage this year or next.

In the meantime, some local governments seem to see the value of ingredient labeling. Joliet, Illinois is the first city with a law requiring restaurants to post a sign disclosing the types of fat they use for frying foods. The sign will list saturated and polyunsaturated fats. This law will pressure restaurateurs to use polyunsaturated fats.

Q: CSPI has monitored manufacturers' health claims, which link their products with disease prevention or a more healthy life. How do you view the current status of health claims and the government's regulation of them?

A: We have mixed feelings about the use of health claims in food advertising and labeling.

On the one hand, we like to see companies advertising the health benefits of their products. We would like more healthful foods marketed more heavily. On the other hand, we fear the companies may be dishonest in their advertising.

Our basis for that fear is the work we have done on food advertising over the past three years. We have seen ad after ad from major companies containing misleading and dishonest information. Examples include marketing a white bread as being as nutritious as whole wheat bread, beef giving strength, which really results from exercise, or coffee as a product that calms you down.

We are concerned that companies will start saying that their foods will make your skin healthy or prevent heart disease, osteoporosis or cancer. These kinds of claims could create confusion in the marketplace.

So we have not endorsed health claims but believe the government should restrict the claims. For instance, we have urged that the claims be restricted to particular nutrients or diseases: fat and heart disease, sodium and high blood pressure. The claims should be limited to a few areas where there is a real consensus in the scientific and medical community.

We have urged the government to develop boilerplate language for the claims. Then a company's advertising people would not invent the wording. They would refer to the government standard.

We have stressed that the specific claim be put in the context of an overall diet. Then you are not just focusing on one particular issue; you are trying to educate people more generally about nutrition....

EXCERPTED FROM

Making It Happen:
A Manual for Local Safe Food Organizers

AMERICANS FOR SAFE FOOD (ASF), PP. 5-10.

Negotiating With Supermarkets

Members of your coalition should contact the consumer affairs departments or store managers of your local supermarkets or grocery stores. Ask for a meeting and offer to send a package of ASF materials with your coalition name and address on them. Enclose your own materials, if available. In any case, you will want to tell them what local groups are represented in your coalition, and approximately how many local citizens you represent.

We recommend that your first approaches to these store representatives be cooperative and educational, rather than adversarial. These stores are basically neutral about offering contaminant-free or organic food. They stand to make a profit if these higher-priced items sell well, and need to be convinced that there is sufficient demand to experiment with certain items in their stores. We are not asking stores to replace food they stock traditionally, but rather that they offer shoppers alternative foods alongside the others.

Store representatives may believe that the supply of organic food is insufficient for them to rely on . This is probably true for certain items. For others, such as beef and chicken, the supply may be ample. Point out that the demand for safe food is growing steadily all across the country, and that inquiries from supermarkets about supplies of particular organic foods will help stimulate farmers to supply such food in the future if they can't already. Suggest that they start by carrying organically grown food in just a few stores.

Tell store officials that industry leaders are moving toward safe food. Inform them that Grand Union supermarkets (an East Coast chain of more than 300 stores) are marketing "natural beef" and are test-marketing "natural chicken." Bread and Circus, a chain of four Boston supermarkets, has built up a thriving business selling "safe food," as have many other natural food stores around the country. Safeway's 125 stores in the United Kingdom offer a sizable array of organically grown fruits and vegetables. These examples indicate a marketing trend fueled by a growing interest in staying healthy. With this sort of presentation, your meeting does not have to be confrontational at all. Quite the contrary – you are encouraging the store to jump on a bandwagon that will improve its image as well as its business in the long run.

You then could discuss ways your coalition could help with the promotion. Many of the coalition's member groups have newsletters, which could feature the store's cooperation and describe what items will be available. You can give an award to the store for its concern for public health and receptivity to consumers. On a national level, ASF has honored 18 corporations, state governments, and individuals as "Safe-food Trailblazers" for their promotion of safer food. You could also offer to send out a press release announcing the availability of certain contaminant-free foods and praising the store for responding to consumer concerns. You can call the local television and radio talk shows and ask them to do a segment on this new development. You can offer to write a letter to the editor of your local newspaper, or ask its editorial page editors to praise the store's progressive stance. Just be sure to wait until the store has actually made a significant change for the better. Stores are very concerned about their image in the community, and you can convince them that carrying contaminant-free food will improve that image.

If the consumer affairs rep. or manager is responsive, you may need to meet with the store's buyers. Most store buyers have never been in contact with producers of organic or pesticide-free food and may not be inclined to seek them out. You may have to seek out national and local producers who can provide the volume of goods that your particular store requires. If local sources exist, offer to set up a meeting with these producers and supermarket representatives so the producers can show off their products and talk about their ability to deliver, their prices, etc.

What about stores whose representatives are uncooperative after you have clearly stated your case? Try going over their heads to the store's owner or home office. Some big chains may have regional offices. If those attempts fail, call or write to us here at ASF headquarters in Washington, and we'll call you to discuss your next move. One possibility is picketing the store, not so much to attack the store as to educate consumers about problems with food. In addition to carrying signs, your volunteers could dress up as pesticide-resistant insects or giant pesticide molecules to get attention. Hand flyers to shoppers on their way into the store.

Certification

"How do I know that the organic food I buy is really grown as promised?" This is a natural question for the consumer concerned about safe food, and as leaders of a coalition promoting organic food, you will have to become experts on certification. Getting to know the people connected with the nearest certifying agencies is also important, because they will be natural allies and supporters for all of the coalition's activities.

Organic farmers have recognized the need for a respected verification process that can offer consumers assurance that food labeled organic meets certain standards. They have responded by forming independent certifying agencies. These agencies add to the credibility of the true organic farmer and increase the integrity of the marketplace in general.

Most of these certifying organizations offer technical assistance to farmers as well. Some hire third-party inspectors to prevent any mischief in the certification process. When a product is labeled "certified organic," it indicates that the farm where the food was grown has met certain standards of organic farming. Generally, an inspector visits each farm that applies for certification to examine the growing process.

"Organic" farming is generally defined as farming that minimizes or avoids the use of synthetic fertilizers, pesticides (including herbicides, insecticides, and fungicides), growth regulators, and feed additives. This is a limited definition, as many certifying agencies have more extensive criteria that a farmer must meet to be certified organic.

There are five primary methods used to verify organic products: questionnaires, affidavits, inspection reports, laboratory analyses, and audits. The following issues can be used to evaluate a particular certification process:

• Testing and Inspection

Most certifiers subject applicants to a soil test or analysis. Some do residue testing as well. You should ask about ongoing inspections, i.e., how often does the certifying agency do spot checks on farms certified organic? Are all inspections made by a third party to assure objectivity?

• Application Process

Most certifiers require farmers to submit an application to be considered for certification. Some require a sworn and notarized affidavit for the first year or longer. In all cases, the farmer must supply the certifier with accurate information about his past and present farming practices.

• Philosophy

Many certifiers believe strongly in the philosophy of sustainability. Sustainable agriculture generally means a system of ecological soil management practices and low-input farming leading to a self-sustained ecosystem that does not require artificial or chemical treatment. Sustainable farming is not necessarily organic farming, but it is a vast improvement over conventional chemical farming.

• Transitional Farmer

Most certification programs require some period of transition before a farm that has been under conventional management can be certified organic. Certifiers usually require delays of at least one year from the last use of prohibited fertilizers and two to three years from the last use of prohibited pesticides.

Some programs do certify transitional farmers as such, partially to encourage more conventional farmers to make the switch to organic techniques. Other certifiers, however, do not see transitional certifications as benefitting the consumer.

• Working Toward a National Standard

Many people in the organic foods business feel strongly that a common definition for "organic" is badly needed to lend more consistency and uniformity to the term. The Organic Foods Production Association of North America (OFPANA) has adopted national guidelines that amount to a first step in that direction. OFPANA is a trade association consisting of organic producers, retailers, and certifying agencies. It considers itself the certifier of certifiers. It will offer its endorsement to certifying agencies whose standards meet the OFPANA criteria.

Because there is no national standard as yet for the certification of organic food, the methods and standards established by certifying agencies do differ. Many of these agencies have published pamphlets or other documents that describe their certification standards and requirements, and they will be glad to send them to you. Although most of these organizations deal most often with farmers, some have consumer memberships as well. In any event, organic farmers are your natural allies. After all, you are the crucial link – your coalition is formed to promote their food! Their organizations should be valuable sources of information, energy, and support for your efforts, and in return your efforts should include a search for helpful information on organic farming or on any grants or support programs available to farmers considering the switch to organic farming.

Resources (for locating sources of organic food)

The following are some of more than 40 certifying agencies in the United States. The list also includes several organizations that can help you locate wholesalers and retailers of organic products.

California Agrarian Action Project
P.O. Box 464
Davis, CA 95617
(916) 756-8518
This group publishes a directory, indexed by commodity for quick reference, of organic food wholesalers, farm suppliers, and other resources.

California Certified Organic Farmers (CCOF)
P.O. Box 8136
Santa Cruz, CA 95061
(408) 423-2263

Demeter Association
West of the Mississippi:
4214 National Avenue
Burbank, CA 91505
(818) 363-7312

Demeter Association
East of the Mississippi:
P.O Box 6606
Ithaca, NY 12851

Farm Verified Organic Program (FVO)
Mercantile Development Inc.
274 Riverside Avenue
P.O. Box 2747
Westport, CT 06880
(203) 226-7803

Maine Organic Farmers' & Gardeners' Association
Jay Adams
P.O. Box 2176
Augusta, ME 04330
(207) 622-3118

Natural Organic Farmers Association (NOFA)
P.O. Box 335
Antrim, NH 03440
(603) 588-6668

NOFA - Vermont
15 Barre St.
Montpelier, VT 05602
(802) 223-7222

NOFA - Massachusetts
21 Great Plain Avenue
Wellesley, MA 02181

NOFA - Connecticut
100 Rose Hills Road
Branford, CT 06405

NOFA - New York
P.O. Box 454
Ithaca, NY 14851

Ohio Ecological Food and Farm Association
7300 Bagley Road
Mt. Perry, OH 43769
(614) 448-3951

Organic Crop Improvement Association (OCIA)
125 West Seventh St.
Wind Gap, PA 18091
(215) 863-6700

Organic Food Network
c/o American Fruitarian Society
6600 Burleson Road;
P.O. Box 17128
Austin, TX 78760-7128
(512) 385-2841
This group publishes a directory of organic food growers and suppliers.

Organic Foods Production Association of North America (OFPANA)
c/o Judith Gillan
P.O. Box 31
Belchertown, MA 01007
(413) 323-6821
OFPANA is a trade association made up of businesses, organizations, and individuals involved in the organic food industry.

Organic Growers and Buyers Association
P.O. Box 9747
Minneapolis, MN 55440
(612) 674-8527

Organic Growers of Michigan
c/o Lewis King
3031 White Creek Road
Kingston, MI 48741
(517) 683-2573

The Organic Network – Eden Acres, Inc.
12100 Lima Center Road
Clinton, MI 49236
(517) 456-4288
This group publishes a directory of organic food suppliers (primarily growers) listed by state and zip code and indexed by commodity.

Tilth Producer's Cooperative
1219 East Sauk Road
Concrete, WA 98237
(206) 853-8449

Sample FOIA Request Letter

Congress enacted the Freedom of Information Act (FOIA) in 1966 to give the public access to information held by the federal government. FOIA gives any person the right to request and receive any document, file or other record in the possession of any agency of the federal government, subject to nine specific exemptions. While the federal FOIA does not apply to state governments, each state has its own laws governing disclosure of records held by state and local government bodies.

Freedom of Information Unit
(Name and Address of Government Agency)

Dear Sir or Madam:
    Re: Freedom of Information Request

Pursuant to the Freedom of Information Act, 5 U.S.C. 552, I hereby request access to (or copy of) all records pertaining to (describe the subject or document containing the information you want).

I am requesting these records (as a representative of the news media, for noncommercial personal use, for an educational institution, etc.)

(If any expenses in excess of $ XX are incurred in connection with this request, please obtain my approval before any such charges are incurred.)

or

(I request a waiver of fees because my interest in the records is not primarily commercial and disclosure of the information will contribute significantly to public understanding of the operations or activities of the government because_____.)

I will expect a response within 10 days as provided by law.

Thank you for your prompt attention to this matter.

Very truly yours,

Your Name

CONSUMER GROUP HONORS "SAFE FOOD TRAILBLAZERS"
AMERICANS FOR SAFE FOOD
Press Release

JANUARY 21, 1987

Americans for Safe Food (ASF), a coalition of over 40 citizens groups, today saluted 18 companies, states, and individuals for increasing the availability of food free of dangerous contaminants.

"These leaders have taken bold steps to rid our food supply of pesticides, hormones, antibiotics, or other dangerous substances," said ASF coordinator Deborah Schechter.

At a press conference in Washington, the following Safe Food Trailblazers were named:

1. **H.J. Heinz Co.,** for making its baby foods only from produce that has not been treated with any of 12 pesticides labeled potentially hazardous by the Environmental Protection Agency.

2. **Grand Union** supermarket chain and Coleman Ranch for spurring the production and marketing of meat virtually free of drug and pesticide residues.

3. The states of **California, Maine, Massachu-** setts, **Minnesota, Montana, Nebraska, Oregon,** and **Washington** for passing laws that define "organic" for labeling purposes, thus promoting consumer confidence in organically-grown products.

4. The **Natural Resources Defence Council,** for providing consumers with valuable information about pesticide residues in imported and domestic food, and for seeking to ban the routine use in animals of antibiotics that are invaluable medicines for humans.

5. **Bread and Circus,** a chain of four Boston supermarkets, for its aggressive marketing campaign that publicized the problems of food contaminants, and for successfully selling food grown without the use of pesticides and drugs.

6. **Rex Oberhelman,** a Minnesota farmer who turned the bankruptcy of his conventional farm into a thriving network of over 50 farms supplying pesticide-free food to dozens of retail stores and restaurants

in five states.

7. The **Rodale Research Center,** which, under the guidance of Robert Rodale, has developed, tested, and publicized new farming techniques that reduce the need for insecticides, herbicides and chemical fertilizers.

8. **Ken and Jim Raffel,** the Arby's restaurant franchisees in Portland, Maine, who use as much locally- and organically-grown food as they can in soups, salads, and other foods.

9. **Dr. Joseph A. Settepani,** a career scientist at the Center for Veterinary Medicine of the U.S. Food and Drug Administration, who has been a consistent advocate for greater consumer protection from potentially dangerous animal drugs.

10. **Safeway's United Kingdom division,** whose 125 outlets offer a sizable array of organically-grown fruits and vegetables; Safeway should provide similar foods to its American customers.

Americans for Safe Food's goal is to convert consumer dismay about health risks in food into the general availability of contaminant-free food.

Americans for Safe Food is coordinated by the Center for Science in the Public Interest, a Washington-based health advocacy group that is supported by its 80,000 members and foundation grants. The ASF coalition includes dozens of organizations ranging from the Texas Center for Rural Studies and the Ohio Public Interest Campaign, to the Environmental Defense Fund and the Consumer Federation of America.◆

A Letter from Cesar Chavez

A contemporary farm worker-consumer national campaign against contaminated grapes as described by Cesar Chavez, leader of the United Farm Workers of America.

Dear Friend,

When I was a teenager, just after the Second World War, I had an experience which changed my life. As you read this letter, I hope it will change yours, too.

It was a typical day of winter farm work in California's Santa Clara valley. My brother and I were working as field hands in an apricot orchard, when one day the owner handed us a bucket of strong-smelling liquid, and told us to put a little around each tree.

We started to pour the stuff, and immediately I began to feel nauseated. Since it was right after lunch, I thought it was something I'd eaten, so I walked back to our car. As I walked, I got dizzier and dizzier, and I passed out for several hours.

When I talked to other farm workers, I found out they'd had the same experience. Farm owners and foremen were telling us it was "medicine" we were pouring and spraying on trees, cotton fields and vineyards. But that "medicine" was giving us skin rashes, making us vomit, causing nosebleeds, and making us weak.

Many things have changed during the past forty years. But there is one aspect of farm work which has, if anything, gotten worse: the use of powerful, danger ous, toxic pesticides on every fruit and vegetable that you and I eat.

And there is conclusive evidence that you don't have to be a farm worker to be affected... *seriously* affected... by chemicals used in agriculture. *All you have to do is eat!*

This means that you and I are subjected to unknown health risks with every meal. And we're subjecting our children, and generations unborn, to those same risks.

I made a vow, as that teenager sickened by a bucket of tree "medicine," that I would make sure, some day, that neither I nor anyone else would ever again have to risk serious illness through the essential work of providing food.

The United Farm Workers of America, which I helped to organize over twenty years ago, is more than ever committed to achieving that goal: a poison-free environment for all agricultural workers... and for you and your own family.

I'm writing to you because *we're in the midst of a campaign* which, if successful, will represent a major breakthrough in forcing agribusiness to stop poisoning workers... and consumers.

Our tactic is simple. We're asking people to not buy table grapes until the growers agree to three demands:

• The *elimination* of five cancer-causing chemicals from all grape fields...

• A *joint testing program* for poisonous substances in grapes sold in stores...

• *Free and fair elections* for farm workers, and good faith collective bargaining in the grape industry.

We know that, with your help, this grape boycott can work. In fact, it worked before; as you may remember, in 1970 we were successful in eliminating such deadly poisons as DDT, DDE, and Dieldrin from every field under United Farm Workers union contract ... thanks to a nationwide grape boycott.

Since then, however, the situation has grown even more serious. Thousands of farm workers are poisoned each year in the grape fields. Testing has identified residues of *more than fifty chemical products on grapes sold to you in stores.*

We've identified the five most lethal substances used in the growing of table grapes. Each of these deadly poisons has a record of injuring – and in some cases killing – farm workers. And each of these deadly pesticides can be present, as a residue, on the table grapes you buy...

• *Methyl Bromide...* extremely poisonous to all forms of life, this fumigant has been responsible for more occupationally-related deaths than any other pesticide. Even non-fatal exposure can cause sever, irreversibly effects on the nervous system, with permanent brain damage, or blindness.

• *Parathion* and *Phosdrin...* can be rapidly fatal, producing illnesses in workers in as little as 20 minutes. Usually sprayed aerially, these poisons cause populations surrounding agricultural areas the same problems as they cause farmworkers, since as much as 90 percent of aerially sprayed pesticides miss their target areas.

• *Dinoseb...* poisoning at first resembles heatstroke, then cumulative doses cause extensive illnesses, including loss of vision. It is much too toxic to be used safely (so poisonous, the EPA has finally banned its use, "pending industry reaction.")

• *Captan...* 344,000 pounds are used annually on table grapes, and residue of this compound is the most frequently discovered material on grapes in stores. Not only can Captan cause cancer, it also causes birth defects and changes in body cells. It is structurally similar to Thalidomide, which caused thousands of babies in Europe to be born without arms and legs.

Not only is each of these deadly pesticides used extensively on the grapes you and I find in stores all across the country. They also have one other thing in common: *each has been recommended for banning by state and federal agencies ... yet they continue in use!*

It's part of a deadly mass of pesticide poisons used annually in grape production... eight million pounds each year of more than 130 chemical compounds.

The most immediate victims of these pesticides are, of course, farm workers. Recent studies have shown that...

• Seventy-eight percent of Texas farm workers surveyed had chronic skin rashes; 56 percent had kidney and liver abnormalities; and 54 percent suffered from chest cavity problems...

• More than 300,000 farm workers are made ill every year through pesticide exposure, with pesticide poisoning incidents doubling over the past ten years...

• The miscarriage rate for female farm workers is seven times the national average...

• Disability days associated with pesticide illnesses have increased by 53 percent since 1979; hospital days by 61 percent...

But *you don't have to be a farm worker to be affected* by the tons and tons of pesticides used in agricultural production... especially in grape production, which is responsible for over one half of all pesticide-related illnesses.

As a consumer, you should know that...

• A recent independent sampling of fruits and vegetables in a San Francisco supermarket showed that 44 percent contained measurable residues of pesticides... 42 percent of those contained more than one chemical – and some had as many as four.

• There is no consistent Federal or state monitoring program for pesticides in food... and such studies as do take place are usually kept secret, so as not to "alarm" the public...

• Pesticides are now thought responsible for groundwater contamination in 23 states... and groundwater provides 50 percent of our country's drinking water supply.

Outrageous? Of course it is. And I can't blame you if, like most people, you've felt helpless to do anything about it.

But now, *there is something you can do!* Something so powerful and effective that if you do it along with us, together we can get those poisons out of our food... permanently!

The first step is simple. **Stop buying table grapes!** And if you see them in a friend's house, talk about what you've read in this letter: the poisons used in growing grapes... their serious health effects... and the United Farm Workers' boycott.

Then you can further help our effort to get poisons out of the fields... and food... by supporting our organizers in their work all across the country.

We consider this campaign so important that we have dispatched our entire union leadership to key cities coast to coast... not just for a few days, but for as long as it takes to make consumers, stores, and distributors aware that we mean business.

I'm sure I don't have to tell you that farm workers are the poorest workers in America. Our members not only do the most dangerous work in the country… they also receive the lowest wages. Therefore, we have no money to spare for even such vital activities as organizing this nationwide grape boycott.

Farm workers and their families are leaving their homes to travel to faraway cities to tell the story of the grape boycott. A donation of as little as $20 will support a farmworker boycotter for a day. For $25, we can print enough materials in our own print shop to leaflet an entire neighborhood. Your $35 donation means food and travel for a volunteer for one week… $100 will provide housing for a farm worker family for one week.

Such amounts are little enough to pay, I think you'll agree, to assure our peace of mind about what we eat. And little enough to pay so that farm workers will no longer have to suffer… as I myself first saw in those apricot orchards 40 years ago… poisoning, deformities, and tragically premature deaths.

Now, it's up to you. For our future… and for yours! For our children… and for your children… please boycott grapes. And send as generous a donation as you can afford, today.

Sincerely,

Cesar Chavez

P.S. I am not asking you to give up wine or raisins. Nor am I asking you to support a large administrative budget because all of us on the union's staff, including myself, still receive only basic subsistence and $10 per week. Can we count on you to support pesticide-free food by sending as generous a donation as you can afford… today?

P.P.S. California Grapes are now being harvested! They will be in the stores continuously from early June until late December. ◆

You can help Cesar Chavez fight against toxic pesticides which threaten farm workers and consumers. Send contributions to:

Cesar Chavez
P.O. Box 62
Keene, CA 93531-0062

Make checks payable to: United Farm Workers.

SECTION 6
A Brief Global View

A broader frame of reference always helps to integrate a stream of facts into a more coherent understanding of causes and effects, of broader benefits and costs, and of wider possibilities for good food for all within a self-sustaining ecology. A few readings here can whet your appetite for books such as *Diet for a Small Planet* by Frances Moore Lappe.

And speaking of more followup, if you wish, peruse the **Resources List** compiled at the end of this volume which refers you to **31 free publications**, two toll-free hotlines and other books available at your library or bookstore.

Together, we can make a difference!

How Our Food Choices Affect the World
BY RONALD E. KOTZSCH
EAST WEST JOURNAL, JUNE 1985, PP. 58+.

Sarah J. is a bright young lawyer with an interest, both personal and professional, in ecological and wilderness issues. She works for an environmental organization at far less pay than she would receive elsewhere. On a typical day she prepares a complaint against a factory that is polluting a local river, solicits donations for the World Wildlife Fund during her lunch hour, and before leaving work, sends a contribution to Oxfam to help with famine relief in Africa.

On her way home, Sarah stops at the supermarket. She buys cereal, milk, eggs, bananas, coffee, cheese, and whole wheat bread. Though she eats little meat, she buys some inexpensive chopped beef for her dog. She notices some apples labeled "local-organically grown," but decides they are too expensive and too spotted. Sarah chooses several cosmetically perfect apples instead, and pops them into a polyethylene produce bag.

Sarah arrives home tired, but satisfied she has made a contribution to the present and future well-being of the planet. And indeed, through her work, her volunteer activity, and her financial contribution she has done so. But she has affected the world negatively as well, in ways she did not intend and does not suspect. Unwittingly, she has supported the exploitation of countless animals, the pollution of the waterways of the nation and the erosion of its soil, the destruction of the tropical rain forest in Central America, and the continuation of lives of privation and hunger for millions of people in the Third World. She has done so in complete innocence, by her purchases in that great unrecognized arena of moral choice – the local supermarket.

Almost 15 years ago, Frances Moore Lappe pointed out that our dietary choices are not purely personal ones, that they affect not only ourselves but our society and the whole planet. In her landmark book, Diet for a Small Planet, Lappe focused on the fact that Americans, because of their meat-based diet, consume an inordinate portion of global food resources. Suggesting that this regime contributes to food scarcity around the world, Lappe encouraged people to adopt a low-meat diet based on grains, seeds, nuts, vegetables, fish, and dairy foods. Many followed her suggestions.

Lappe and others have since observed that our dietary choices affect not only the global food supply, they also influence environmental, social, economic, and even political conditions around the world. When we purchase a particular food, we give our tacit approval and support to those who produced it and brought it to the supermarket shelf. As Lappe puts it, "Every decision we make about food is a vote for the kind of world we want to live in." Yet even people with genuine humanitarian and ecological concern seldom have a clear idea of the impact of their dietary choices. A look at some of the foods we Americans commonly use, and at the kind of world we vote for when we buy them, is instructive and perhaps surprising.

Meat:

Eating meat involves the killing of animals, a fact of which many meat-eaters are often unconscious. A single visit to a slaughterhouse has been enough to convert many to vegetarianism. But even if we have no qualms about the slaughter of animals for food, there are serious environmental and other ethical issues involved.

The grazing of cattle, sheep, and goats for meat can have a valid place in an ecologically balanced food economy. For one thing, not all land is fit for the cultivation of food crops such as grains and beans; the soil may be poor and there may be inadequate water. In such cases animals change inedible forage plants into human food. Attempts to develop agriculture on such lands can prove disastrous. In recent years, the cultivation of wheat, on range lands in Colorado, for example, has led to the destruction of the topsoil.

However, the number of animals raised exclusively on pastoral lands in America is small. Most spend a good part of their lives in feed lots. These are enclosed areas where thousands of animals are crowded together and fattened with a diet of corn, soybeans, and other potential human foods. Lappe's original point that we use human food to feed livestock and thus deprive hungry people is still a valid one. And we are doing this in what amounts to animal concentration camps. Life in these cattle feed lots is so unhealthy that the animals are constantly dosed with antibiotics. Half of the antibiotics produced in this country are used in livestock feeds.

The animals, of course, produce manure – literally mountains of it. Whereas if properly treated, it might be used as fertilizer or even as a source of methane fuel, usually it is treated as a waste product. It is allowed to wash away into streams, rivers, and lakes where its high concentrations of nitrates and phosphates upset the ecology of these waters. The Chesapeake Bay, for example, long a major source of fish and shellfish, has become almost barren. Animal waste run-off from farms along its source rivers has virtually destroyed the balance of plant and animal life.

If we buy beef, especially the cheaper varieties used in fast-food outlets, there is another fact to consider. A substantial amount of this beef comes from Central American countries where the cattle industry has been involved in the wholesale destruction of the tropical rain forests. Huge tracts of forest are cleared by bulldozers and are sown with grazing grasses. Cattle are pastured there and the meat sold at low prices to American concerns. In a few years the fragile topsoil is depleted and the area abandoned. The band of massive rain forests that gird the earth's equator is a key factor in the ecology and climate of the planet. It is being devastated in the noble cause of the cheap hamburger.

Dairy Foods:

Like many total or semi-vegetarians, Sarah believes that she is doing cows and other livestock a good turn by not eating their meat, and by substituting cheese, yogurt, and milk in her diet. And to be sure, she is refusing to underwrite the slaughter and the raising for slaughter of these animals. But to the compassionate observer, many dairy farms have little to recommend them over the feed lot and the slaughterhouse.

In a former, more innocent age, milk cows led a quiet, bucolic life. Ole Bessie roamed hillside pastures and twice a day was hand-milked in the barn by her owner. She performed a useful service changing weeds and forage into human food and providing fertilizer.

Today, this Sound of Music idyll is largely a thing of the past. Very likely Ole Bessie is chained for most of her life in a huge barn with several hundred other cows. She can do little more than eat, lactate, and defecate. Her incarceration makes her prone to infection, so she receives antibiotics as well. At milking time vacuum hoses are attached to her teats and her udders are pumped dry. Like her brothers and sisters in the feed lot, Bessie is treated as little more than a biological machine. As soon as her milk production falls below a certain level, she is sent to the slaughterhouse.

Clearly, the use of dairy foods also raises serious issues for the humane and environmentally aware consumer. Bessie may not be the smartest of God's creatures, but she probably knows the difference between life in a stall and in a pasture. With our glass of milk, bit of cheese, even our wholesome bowl of yogurt, we are contributing to her life of incarceration. And we are contributing as well to the relatively inefficient use of human foodstuffs.

Eggs:

There is something nostalgic and appealing about a flock of hens, clucking about the yard, trysting with the rooster behind the willow tree and laying an occasional egg to nourish their human caretakers. But again this pleasant image is a thing of the past. Most chickens (both egg-layers and those destined for the frying pan) are raised in conditions that make a feed lot or dairy barn look like four-star luxury accommodations.

The typical chicken spends her life in a windowless "factory," in a cage one foot high, wide, and long. She barely has room to stand up, let alone move around: On that same spot she eats, defecates, and lays her eggs. She is fed hormones and doused with antibiotics. The factory lighting is manipulated so the "day" passes quickly and the chicken lays more frequently. The bird is lucky to survive a single year under such conditions.

"Luxury" Foods:

A substantial part of the American diet consists of "luxury" items. These include sugar, coffee, cocoa, tea, and tropical fruits such as bananas. These have become such an accepted part of our daily life that we scarcely consider them "luxuries." Yet they are. These foods are not necessary to human life and until relatively recently were unknown or little used in most of Europe and America. They are also luxuries in that their production exacts a toll on both the land and the people involved in their production.

From their first contacts with the lands of Asia, Africa, and South and Central America, the European nations recognized a great agricultural resource. Especially in the nineteenth century, the colonial powers, including the United States, developed this resource to meet their own needs by turning these areas into suppliers of cheap agricultural products. These items included rubber, cotton, and hemp as well as food items such as coffee, sugar, and bananas.

In some regions plantations were immediately established to raise these products on a massive scale. In others, taxation forced small farmers to cultivate export cash crops, and eventually, in hard times, to sell off their land. In most areas the net result was the same. The agricultural land was concentrated in the hands of a few native or foreign concerns, and most of it was dedicated to the monoculture of the export cash crops. Land used for domestic crops decreased. Many peasants, deprived of even a garden plot, became dependent for survival on imported foods, which could be

---

Pollution and Food Processing

Food processing in an industrialized country often means costly, polluting factories (see table) that employ little of the work force. These factories consume much energy, recyclable raw materials, and water. The concentration of factories, furthermore, necessitates expensive transportation and infrastructural supports.

It is no wonder processed food costs more. Increased costs are the result of preparation, packaging, and other processes which nutritionally deplete the final product more than they add to its value.

Water Pollution Resulting from Food-Processing Industries

AGRICULTURAL AND FOOD-PROCESSING INDUSTRIES	SOURCES OF THE PRINCIPAL WASTES	CHARACTERISTICS
Canned fruits and vegetables	Cleaning, squeezing, blanching, and pressure cooking of fruits and vegetables	High MS content; gels and organic material dissolved; pH sometimes alkaline; starch
Meat processing and salting	Sheepfolds, slaughterhouses, treatment of entrails, condensers, fats, and water from washing	High concentration in dissolved organic material and in suspension (blood, proteins), fats, NaCl
Food for livestock	By-products from centrifuge treatment, pressing, evaporation and residue from washing	Very high DBO,* only organic materials, odor solvents
Dairies	Dilutions from whole milk, skim milk, buttermilk and serum	Very high concentration of dissolved organic materials, mainly protein, lactose, fats
Sugar refineries	Washing and transportation of sugar beets: diffusion, removal of scum, condensed evaporation, regeneration of ion exchangers	Very high concentration in dissolved organic matter and matter in suspension (sugars and protein).

* Demande biologique en oxygene = biological oxygen demand
Excerpted from: Degremont, Ste, "Memento technique de l'eau," Rueil-Malmaison.

acquired only with cash earned on the plantations. Native people became indentured through the powerful agricultural interests, and bound to a life of poverty and hunger.

Although the colonial era has passed, and most of these countries are independent, little has changed. In the "banana republics" of Central America, in the coffee-growing nations of sub-Saharan Africa, in the sugar-producing areas of the Philippines, the agricultural, social, and economic situation is essentially the same. Most of the land is used for cash crops and is controlled by a native elite or by multinational corporations. In El Salvador, for example, 81 percent of the arable land is devoted to coffee cultivation and is owned by three percent of the population. The main beneficiaries of this economy are the urban middle and upper classes. They enjoy a North American lifestyle replete with cars, TV's and refrigerators, while the people who work the land live at or below subsistence level.

In some areas the issue is not merely one of social and economic justice, but of survival. According to *Food First*, a book on world hunger by Lappe and Joseph Collins, cash crop agriculture has been an important factor in the current famine in sub-Saharan Africa. Traditional food crop farming protected the soil and produced a surplus that fed the people during drought periods. Throughout this century, European concerns practicing large scale cultivation of cash crops such as peanuts and cotton reduced the amount of land in local food production. They cut down the few precious trees of the savannah to permit mechanized farming. They "mixed" the soil and when the land was no longer fertile they abandoned it. Once-productive land became desert, useless for growing food and contributing perhaps to the long drought.

The problems of social and economic justice, of poverty, hunger, and famine in the Third World are, of course, complex ones. But, at the least, we should be aware that our choices in the supermarket are related to them. The banana we eat with our morning cereal and the cup of coffee we enjoy at noon connects us to these realities.

Grain, Beans, Vegetables and Fruit

Most of the agricultural products that Americans consume are domestic products. The issues raised by our use of them, while close to home, are also serious and complex.

Most of the food grown in the United States is produced by standard commercial agricultural techniques. This system involves the use of chemical fertilizers and petroleum-based pesticides and herbicides. Thus, American commercial agriculture is often a major polluter of the environment. Fertilizer runoff is a major cause of water pollution. Broadcast spraying puts agricultural poisons into the air and water as well as the soil.

In addition, this is not a sustainable agriculture. It is steadily destroying the topsoil and the living organisms within it. Some observers say that by the turn of the century much of the farmland now in use will be infertile. We may be approaching another "Dust Bowl" era, with disastrous consequences for ourselves and for the rest of the world.

Almost every time we buy a loaf of bread, a can of beans, or a head of lettuce, we are supporting this system of agriculture. And Sarah, though she may oppose industrial polluters on the one hand, is abetting agricultural polluters on the other through her food purchases. While she is concerned about food scarcity in other parts of the world, she may be contributing to an eventual shortage closer to home.

We have by now painted a depressing picture. When we buy meat or dairy we are supporting exploitation of animals, environmental pollution and the use of human food as "feed." When we stop at Burger King for a quick meal we are helping to swing the axe on the tropical rain forests. When we eat a banana or sip a cup of coffee we may be underwriting an unjust social and economic system and helping to keep people in want and hunger. Even when we buy a loaf of whole wheat bread we are supporting a system of "dead-end" agriculture. What indeed can a humane, concerned, and aware person buy and eat? One is tempted to despair. The hypersensitive spirit resolves to become a "breatharian." The rebellious one throws up his or her hands impatiently and marches down to the local burger joint. "Apres moi le deluge!"

Diet For A Just Planet

How then is it possible to make our food choices congruent with the way we would like the world to be? Is it necessary to become a homesteader or food gatherer in some remote area? Not really, but it does require some energy, vigilance, and integrity. Let's look at each of the groups of foods we considered and at some practical things we can do regarding each of them.

It is possible to live without meat, dairy foods, poultry, and eggs. Vegans (people who eat foods only from plant sources) of past and present have proven this. Certainly, most people could manage with smaller amounts of these items, which are increasingly associated with various health problems, from obesity to heart disease. If we do wish to use them, it is important that we find producers who are ecologically aware as well as humane. There are livestock farmers who range-feed their animals and who use the manure for fertilizer, dairy farmers who allow their cows fresh air and sunshine and do not saturate them with hormones and antibiotics, and chicken farmers whose animals live normal lives out-of-doors.

Products from such sources are available but often are difficult to find. Your local natural foods store may carry some. If so, they will be clearly marked. A "righteous" carton of eggs, for example, will read: "These eggs are from grain-fed, free-ranging, non-medicated chickens." If no such items are stocked, tell the store owner that you are interested in them.

Coffee, sugar, cocoa, and other Third World export crops are not necessary to life. Many societies do quite well without them. If we so choose to use them, again we should explore alternative sources. Since the sources lie on the other side of the world, this is not easy. Happily, some groups have become active in this enterprise.

One such pioneer is a tiny concern in Fort Wayne, Indiana, called "Friends of the Third World" (611 W. Wayne St., Fort Wayne, IN 46802). It sells by mail-order only products from countries in which the local workers control the land and get a fair share of the profits, including coffee and sesame tahini from Nicaragua, coffee and tea from Tanzania, vanilla beans from Madagascar, and cashews and tea from Mozambique.

The vast bulk of grains, beans, nuts, seeds, vegetables, and fruits are produced by regular commercial agriculture. However, there is a small but growing community of ecological farmers around the country who use organic rather than chemical fertilizers and employ natural methods of pest and weed control rather than poisons. Thus, they nurture rather than pollute their land and the larger environment. Theirs is a sustainable agriculture which will keep the soil alive and productive. Most of these farmers operate on a small scale and are struggling to survive. They need and deserve the support of the environmentally aware consumer.

While the number of large family farms is decreasing, especially in the Midwest, the number of small family farms around the country is increasing. Some are ecologically oriented, some are not. In either case, we should support a local producer when we have the opportunity. Energy (oil), truck exhaust, and noise are each involved in the transport of foods from one end of the country to another. If we are concerned about oil exploration in environmentally delicate areas, we should also be concerned about the fact that most of the vegetables and fruits on our dinner table have been trucked in from California.

Grow Thyself

A second option is to grow our own food. Few people are inclined to become self-sufficient home-steaders, but many have the ability to produce at least part of their own food. These products will be organic, environmentally benign, and untainted by social injustice.

Most of us are content to grow a few vegetables in our garden or community plot. According to one California-based group, however, it is possible to produce a sizable portion of our entire needs on a relatively small plot. This idea was advanced by John Jeavons in his book on biointensive techniques, called "How to Grow More Vegetables Than You Ever Thought Possible on Less Land Than You Can Imagine," and has been expanded by Jeavons' associate, David Duhon in "One Circle." (Both books are available from Ecology Action, 5798 Ridgewood Road, Willits, CA 95490.) Duhon maintains that it is possible to produce a nutritionally complete food supply for one person on a circular plot ("one circle") of about 1,000 square feet. He allows that the exact crops and techniques have yet to be perfected; he proposes, though, as a starting point, a list of fourteen basic food crops, including wheat, sweet potatoes, soybeans, sunflowers, parsley, collards, and garlic. Whether Duhon's proposal is viable or not (for one thing, he posits an eight-month growing season), his basic message is clear: we can raise much more of our own food than we usually do.

These measures may also seem futile. What difference will it make if I, one individual, insist on organic sunflower seeds or investigate the pedigree of my apples? It may not make a difference. But then again, it might. The fact is, the destiny of the planet is determined by the daily, seemingly petty decisions of the millions of individuals who comprise humanity and who consume its resources daily. In either case, as Kant observed, the moral person acts always as he or she would want all of humanity to act. This is a high ideal, but there is no better place to start to realize it than in our choice of daily foods.

Ronald E. Kotzsch, Ph.D., is the author of *Macrobiotics Yesterday and Today,* from Japan Publications.

The Cornucopia Papers:
Have You Ever Asked These Questions About Your Food?,
BY ROBERT RODALE
Organic Gardening, October 1980, p. 264.

As I said in the first two articles in this series, the coming food crunch is likely to be a much more serious problem than the energy crisis we have been living with for the past few years. Energy problems are beginning to have a ripple effect through the food system, which is just one small part of the problem. Every day there are more mouths to feed, both here and abroad. At the same time our soil is wasting away at a rate that is becoming frightening. It is possible to extend current trends into the foreseeable future and see an America with little topsoil left, and many millions of people unable to pay the high prices asked for food of decent quality.

There is a bright side to this situation, though. Tremendous potential for increased production of food remains – if we use more of the idle land around our homes and cities for intensive production of vegetables, fruit, beans and many other types of food for people. (Horticulture is far more productive of food than is agriculture, per unit of land area.) Also, by using land close to where we live, the expensive and wasteful practice of shipping food long distances will be reduced.

The Cornucopia Project of Rodale Press is an effort to collect information about the ways the U.S. food system needs to be restructured to prevent a food crisis. We think that if people have a chance to look at the big picture that is formed by all the elements that constitute our food system, they will be able to see both the size of the problem and a road leading to a solution. As long as we just think about our own little corner of the system, without trying to see how everything fits together, the present decline of our food production base will not only continue, but will accelerate.

The best way to know how vulnerable you are to future food-cost inflation, and also to know what you can do about it, is to ask yourself more probing questions about your food. These are the traditional questions people usually ask themselves:

"What am I hungry for?"

"What is nourishing for me?"

"What will it cost?"

Those are very important questions, and I don't mean to downgrade them. But go deeper! Try to ask yourself additional things that will tell you how dependent you are on a food system that may not be able to function in the old way much longer. And think of the information you will need to make your food system more reliable and more self-sustaining.

We may be able to help. Here are some probing questions about food that have been drawn up by researchers at The Cornucopia Project. If they seem to you to be easy and you can answer them well, then you have an important depth of knowledge that can be a useful tool in working for greater food security. If you draw a blank on most of these questions, then you need to get to work quickly to expand your knowledge.

Where Does Your Food Come From?

Think of this question in an item-by-item way, as well as generally. How much comes from your garden, and how much do you buy? Do you have any idea where the lettuce you buy is grown, or where your meat was produced? What about cheese, bread, flour, potatoes and other staples? Try to get a handle on where they originate and whether you could get any of them from sources closer to your home.

Do You Know How Much Your Family Eats In a Year?

The importance of food becomes more apparent when we get an idea of the quantity used over a period of time. The average American eats 1,451 pounds of food a year. That breaks down to 371 pounds of fruits and vegetables, 143 pounds of cereals, 239 pounds of meat and fish, 353 pounds of dairy products, plus 345 pounds of everything else. Multiply that by 230 million people in the U.S. and you can see how important is the system that keeps food coming to us regularly.

Do You Know How Much You Produce in Your Garden?

Have you kept track of what you've grown, how much, and what its value is? Last year, according to the Gallup survey sponsored by Gardens For All, 33 million American households (42 percent) grew some of their own food in home or community gardens. The total value of produce from all our gardens in 1979 was $13 billion–an average yield of $386 for each garden.

With an average cost of only $19, the total saving was $367 for every family growing food in a typical garden of only 600 square feet–or a plot approximately 25 feet on each side.

Record-keeping is the basis of productive planning. If you know how much food you are growing, you can tell how effective your methods are, and what impact fertilizer, soil improvement, weather and improved tillage have. Good records don't have to be complex or time-consuming to collect. And you'll find that they become very useful, over a period of years, to tell what is happening to your garden's fertility.

Do You Know How Much You Could Produce?

What would happen to your annual production rate if you enlarged your garden, put in more time, or grew varieties that produce more in a smaller space? What if you used home food systems in addition to gardening, such as small-scale fish production, making tofu and other foods from soybeans, grinding your own flour, and so forth? What percentage of your total food needs could you handle that way, and what would be the impact on your budget?

How Large Is Your Food Reserve?

Suppose your normal sources of food were cut off for any reason. The cause could be a natural disaster, an economic crash, or panic buying for a variety of reasons. How long could you keep going on the food you have on hand? We now see the importance of a national security reserve of oil (which has not yet been created). A food reserve is even more basic, and should be localized to be most effective. The worldwide store of grain is down to about a 50-day supply. How do you rate in food security by comparison?

Do You Know Where The Nearest Farm Is, And What Food You Could Get From It?

Today, many farmers buy almost all their food in stores. Farming has become specialized, with even family farms producing only one or two crops which are processed into food at distant locations. That single-cropping and specialization is a big cause of the vulnerability of our food system. Do you have any balanced, diversified farms near you? Can you buy local produce at roadside stands?

How Much Food Could Your Community Produce?

Garden potential counts in trying to answer that question, but so do a host of small-farming options. Much vacant land near towns and cities could be used for intensive food production if people saw the need, and if economic factors favored regional food production. An inventory of local food-production potential is an important first step toward planning a food system which would operate with less reliance on high-cost energy, and which could protect regions against inflation and shortages. Are there potential farm/garden lands in your area that are currently idle?

Do You Know How Much Soil Is Lost In the Production of Your Food?

Each pound of corn grown in Iowa is matched by the erosion of six pounds of topsoil. Wheat can be even more destructive. Twenty pounds of topsoil are lost for every pound of wheat produced in eastern Washington State. Most other farm crops cause unacceptable losses of soil as well. When you eat, do you ever think about how much irreplaceable soil is being lost in the process of bringing that food to your table? Are there ways that these losses can be reduced, or even reversed?

What Is The Energy Cost Of Your Food?

Very little energy is needed to produce a potato in your garden or on a nearby farm. But a box of instant potato flakes requires nearly 10 times as much energy as to produce the same amount of raw potatoes. Some foods are as much like a gas-guzzler as any overstuffed old-style car, while others are as lean and efficient as a front-wheel drive subcompact. Do you know which category different foods are in?

Who Is In Control Of Our Food System?

As I pointed out last month, you are in almost total control of the food you grow in your garden. But food that you buy can be under the control of organizations or even countries you know little about. Consumers who know little about the control of their food supply can't do much to change things. Why do some items suddenly jump in price, while others hold steady or even drop? Who decides what new products will appear in stores, or what old ones will disappear?

Do You Know How Government Food Policy Affects You?

Many government programs influence farming, shipping, storage and processing of food. But most consumers have no idea what these policies are or what they are doing to the cost and availability of food. Use of drugs to fatten animals faster can reduce effectiveness of drugs you may need badly when sick, but government policy has yet to stop that dangerous practice. The 1981 farm bill could help improve the food system, but few nonfarmers will try to influence it. How can one individual affect government policy? Do you want to learn more about food policy before a food emergency happens?

Does What You Choose To Eat Make A Differ-

ence?

Think for a moment about the implication of a large number of individual changes in food preferences. Every vote counts at election time, and so do purchases at the food store. Food companies are now planning to introduce and sell more natural and less-processed products, but most companies don't see those foods becoming mainstream items soon. However, if more people switched to more fresh and healthful foods, the impact would be felt up and down the commercial food chain, and policy changes would result. How an we get companies to produce and market more fresh and healthful foods?

Could You Plan to Make Yourself More Self-Reliant?

There are simple steps everyone could take to create more food security. Building a bigger reserve of food is one. Another is to produce more yourself, and do more home processing. That could mean equipping your home with more storage space and tools for processing. Learning to use more whole grains and similar staple foods might be another useful step. Can you make a list of all the things you could do to increase your personal food independence and form them into plan for action?

How Much Self-Reliance In Food Is Possible For The Average Person?

Getting an answer to that question is one of the goals of the Cornucopia Project. In fact, the process by which we try to get an answer illustrates very well how the Cornucopia Project can work. The first step is for you to think through your food situation and see how dependent you are on our food system, as well as how much self-reliance you could achieve. Then tell us, and we will begin to form a picture of the potential for self-reliance nationally. Of course, the purpose of the Cornucopia Project is not limited to promoting home food production, but that could be an important aspect of national food security. You may be able to benefit by knowing more about how much food you can produce, and we can use information about that potential to show how important individual action can be.

Where Is The U.S. Food System Going?

We are all aware of how farms, food stores, and the type of food available have changed. If that kind of change continues into the future, what kind of food will we have, and what will the system of producing and delivering food be like? Is this the kind of food system we want? Think about those questions in terms of your own desire for food, and your community's situation as well.

Where Could The Food System Be Headed?

Are we as individuals trapped in a system of producing food over which we have no control, or is there still time for us to learn what is happening and exert pressure for change? What options for learning and for influence do we have? Is it possible for us to create a plan that would show how the abundant resources of this country could be managed to produce large amounts of good food at reasonable cost and without burning up our soil and fuel resources?

I have only one more question to ask of you. It is this: What do you think of these questions? In attempting to answer that, keep in mind that sometimes asking the right questions can be more important than getting the "right" answers. A good series of probing questions can start people thinking about a taken-for-granted thing like food in new ways. There is more to food than breakfast, lunch and dinner. There is life in our food, and we need to start asking how secure is the source of that life, who controls it, and what we can do to protect it for the future. If we keep asking the right questions, both of ourselves and of the organizations and people who produce food, we can create a consciousness that will lead to constructive action.

You may be able to think of other, better questions that should be asked. And you may feel that some of my questions are more important than others. Our plan is to use your response to this suggested list of questions to create a Cornucopia Agenda of questions that could form the basis for a new kind of analysis of the merits and possibilities for improvement of the U.S. food system. We need your thinking, and your ideas, to make that agenda as effective as possible.

SECTION SEVEN -RESOURCES

FOOD ACTION GROUPS:

While not all of the groups listed below are concerned exclusively with food-related issues, they are all active in at least one of the areas of food safety discussed in this book.

Americans For Safe Food, 1501 16th Street, N.W., Washington, DC 20036

Center for Science in the Public Interest (CSPI). (Americans For Safe Food is a project of CSPI. CSPI is located at the Americans For Safe Food address.)

Clean Water Action Project, 317 Pennsylvania Avenue, S.E., Washington, DC 20003

Community Nutrition Institute, 2001 S Street, N.W., Suite 530, Washington, DC 20009

Environmental Action, 1525 New Hampshire Avenue, N.W., Washington, DC 20036

Food Research and Action Center, Suite 500, 1319 F Street, NW, Washington, DC 20004

Institute for Food and Development Policy, 145 9th Street, San Francisco, CA 94103

National Coalition Against the Misuse of Pesticides (NCAMP), 530 7th Street, S.E., Washington, DC 20003

Natural Resources Defense Council, 122 E. 42nd Street, New York, NY 10168

Public Citizen Health Research Group, 2000 P Street, N.W., Suite 700, Washington, DC 20036

Public Voice for Food and Health Policy, 1001 Connecticut Avenue, N.W., Suite 522, Washington DC 20036

The following citizen groups focus primarily on the issue of food irradiation:

Coalition for Alternatives in Nutrition and Health Care, Box B-12, Richlantown, PA 18955

Consumers United for Food Safety, Box 22988, Seattle, WA 98122

Health and Energy Institute, 236 Massachusetts Avenue, N.E., Suite 506, Washington, DC 20002

National Coalition to Stop Food Irradiation, Box 590488, San Francisco, CA 94159

New York Public Interest Research Group (NYPIRG), 9 Murray Street, New York, NY 10007

People for Responsible Management of Radioactive Wastes, Colleen McGrath, 146 Mills Street, Morristown, NJ 07960

Vermont Alliance to Protect Our Food, Box 237, Vergennes, VT 05491

FREE INFORMATION SOURCES:

From the U.S. General Accounting Office, P.O. Box 6015, Gaithersburg, MD 20877; telephone 202-275-6241:

Federal Regulation of Meat and Poultry Products – Increased Consumer Protection and Efficiencies Needed, GAO/RCED-83-68, May 4, 1983.

Improved Management of Import Meat Inspection Program Needed, GAO/RCED-83-81, June 15, 1983.

Monitoring and Enforcing Food Safety – An Overview of Past Studies, GAO, September 9, 1983.

USDA's Oversight of State Meat and Poultry Inspection Programs Could Be Strengthened, GAO/RCED-84-23, October 21, 1983.The USDA must inspect all meat and poultry products sold for human consumption in interstate and foreign commerce, but states are permitted to inspect intrastate products provided that USDA has certified that their inspection laws and programs are at least equal to those of the Federal Government.

Problems In Protecting Consumers From Illegally Harvested Shellfish (Clams, Mussels, And Oysters), GAO/HRD-84-36, June 14, 1984. FDA has no enforcement authority to ensure adherence to the program's guidelines.

Evaluation of Selected Aspects of FDA's Food Manufacturing Sanitation Inspection Efforts, GAO/HRD-84-65, August 30, 1984.

FDA's Oversight of the 1982 Canned Salmon Recalls, GAO, Sept. 12, 1984.

Legislative Changes and Administrative Improvements Should Be Considered For FDA To Better Protect The Public From Adulterated Food Products, GAO/HRD-84-61, September 26, 1984.

Imported Wines: Identifying and Removing Wines Contaminated With Diethylene Glycol, GAO/RCED-86-112, March 1986.

Nonagricultural Pesticides: Risks and Regulation, GAO/RCED-86-97, April 1986.

Pesticides: Better Sampling and Enforcement Needed on Imported Food, GAO/RCED-86-219, September 1986.

Pesticides: Need to Enhance FDA's Ability to Protect the Public From Illegal Residues, GAO/RCED-87-7, October 1986.

From the Food Safety and Inspection Service (FSIS) Publications Office, Room 1165-South, U.S. Department of Agriculture, Washington, DC 20250:

Safe Food To Go, H & G Bulletin 242, 1985.

The Safe Food Book – Your Kitchen Guide, H & G Bulletin 241, 1984.

Talking About Turkey – How to Buy, Store, Thaw, Stuff and Prepare Your Holiday Bird, H & G Bulletin 243, 1984.

Cooking Meat and Poultry Products (Wall Chart), FSIS-6, 1981. Available in Spanish as FSIS-6-S.

Storing Meat and Poultry Products (Wall Chart), FSIS-7, 1981. Available in Spanish as FSIS-7-S.

Food-Borne Bacterial Poisoning, FSIS-9, 1980.

Safe Handling of Delicatessen Meat and Poultry, unnumbered, undated. A one-page flyer complementing and summarizing the information contained in another FSIS publication, FSIS-17.

Hamburger – Questions and Answers, FSIS-19, revised 1981. Answers to commonly asked questions about purchasing, cooking and storing hamburger.

Know Your Molds, unnumbered, undated. Reprint of a magazine article about good and bad molds, how molds grow, "refrigerator" molds, coping with molds, and the best way to keep food from growing "moldy."

Meat and Poultry Products – A Guide to Content and Labeling Requirements, H & G Bulletin 236, 1981. Lists minimum standards of composition for more than 200 popular meat and poultry products.

Sodium Counting Down, unnumbered, undated. Information on how to cut back on your sodium intake, what to look for when reading labels, and a listing of the sodium content of many foods.

Do Yourself a Flavor, FDA Reprint 84-2192, 1984. Fact sheet listing dozens of herbs and herb blends which can be used as alternatives to salt for flavoring foods. Lists uses for each.

Dietary Guidelines for Americans – Avoid Too Much Sodium, unnumbered, 1986. One of a series of dietary bulletins with tips on where sodium is found in your diet, a guide to the sodium content of foods, estimating the amount of sodium in your diet, and suggestions on how to avoid too much sodium.

FROM THE CONSUMER INFORMATION CENTER:

The Consumer Information Center in Pueblo, Colorado, publishes a Consumer Information Catalog four times per year. Each catalog lists over 200 sources of consumer information available from government agencies for free or for a minimal charge. The catalog is available for free by writing to the Consumer Information Center, Pueblo, CO, 81002. Following is a partial list of free publications available in the catalog which focus on food safety. Please note that if you order more than one free publication from the Consumer Information Center you must pay a $1.00 handling charge for any amount over one (not $1.00 for **each** publication after the first, but one dollar for **all** publications after the first).

Consumer's Guide to Food Labels, 515T, Food and Drug Administration (FDA), 1983.

Diet, Nutrition and Cancer Prevention: The Good News, 517T, National Institutes of Health (NIH), 1987. An estimated one-third of all cancer deaths may be related to what we eat. This booklet will help you select, prepare, and serve healthier foods. Includes list of high-fiber and low-fat foods.

Food Additives, 518T, FDA, 1982. Why chemicals are added to foods and how additive use is regulated.

Meat and Poultry Labels Wrap It Up, 519T, U.S. Department of Agriculture, 1987. Understanding the information on labels for meat and poultry products: low-calorie, low-salt, and low-fat claims; additives; government inspection marks; and more.

A Word About Low-Sodium Diets, 525T, FDA, 1984. Includes recipes for salt substitutes.

TOLL-FREE HOTLINES:

Meat and Poultry Hotline – 1-800-535-4555 (447-3333 in Washington, DC) for questions concerning the safety or wholesomeness of meat and poultry products, including questions concerning your storage and preparation of these food products.

Cancer Information Service – 1-800-4-CANCER for telephone answers to your questions about cancer cause, prevention, diagnosis and treatment. In Alaska call 1-800-638-6070 and in Hawaii, on Oahu, call 524-1234 (call collect from neighboring islands). Assistance for Spanish speakers is available during daytime hours in California, Florida, Georgia, Illinois, northern New Jersey, New York and Texas.

From Americans For Safe Food: Americans For Safe Food (ASF) will send a free copy of their publication, *Making It Happen: A Manual for Local Safe Food Organizers*, to citizens wanting to increase their involvement in the fight for food safety. The ASF address is given above at the beginning of this Resource listing under the Food Action Groups heading.

ADDITIONAL MATERIALS YOU CAN OBTAIN

From Your Library or Bookstore:

Keep It Simple: 30 Minute Meals from Scratch, by Marian Burros, William Morrow and Company, 1981. A potpourri of quick, cheap, nutritious meals.

Jane Brody's Nutrition Book, by Jane Brody, W.W. Norton and Co., 1981. A lifetime guide to good eating for better health and weight control.

The Goldbecks' Guide to Good Food, by Nikki and David Goldbeck, New American Library, 1987. A complete shopping guide to the best, most healthful foods available in supermarkets, natural food stores and by mail.

The Fat of the Land, by Fred Powledge, Simon and Schuster, 1984.

Garden Fresh Cooking, by Judith Benn Hurley, Rodale Press, 1987. (Rodale Press features numerous publications on healthful farming, gardening, cooking, and eating. For more information, contact them at 33 E. Minor, Emmaus, PA 18098; tel. 215-967-5171.)

Retreat From Safety: Reagan's Attack on America's Health, by Joan Claybrook and the staff of Public Citizen, Pantheon Books, 1984.

From the U.S. Government Printing Office, Superintendent of Documents, Washington, DC, 20402:

Adverse Reactions to Foods, S/N 017-044-00045-1, 1984. Discusses current knowledge in the area of food allergy, 117 pp. ($9.50)

Breast Feeding, S/N 017-026-00084-4, 1979, 22 pp. ($3.25)

Eating Hints: Recipes and Tips for Better Nutrition During Cancer Treatment, S/N 017-042-00154, 1980, 86 pp. ($7.50)

Buying Food: A Guide for Calculating Amounts to Buy and Comparing Costs in Household Quantities, S/N 001-000-03811-7, 1978 (revised ed.), 71 pp. ($4.50)

Discovering Vegetables: The Nutrition Education Guidebook for School Food Service Managers and Cooperators, For Use With Children Ages 5 Through 8, S/N 001-024-00201-1, 1975, 16 pp. ($2.75)

Do Your Own Establishment Inspection: A Guide to Self Inspection for the Smaller Food Processor and Warehouse, S/N 017-012-00305-2, 1982, 20 pp. ($3.00)

Eating for Better Health, S/N 001-000-04243-2, 1981, 26 pp. ($3.50).

Gardening for Food and Fun: Yearbook of Agriculture - 1977, S/N 001-000-03679-3, 1977. A complete guide, 392 pp. ($12.00).

From Public Citizen, a national membership organization founded by Ralph Nader, located at 2000 P Street, NW, Washington, DC 20036:

Freedom From Harm: The Civilizing Influence of Health, Safety and Environmental Regulation, by David Bollier and Joan Claybrook, Public Citizen and Democracy Project, 1986, 302 pp. ($10.00)

Public Citizen Health Research Group **Health Letter,** monthly newsletter on health, medical care, drug safety and food safety issues. To receive a one year subscription to the "Health Letter", send $18.00 to Health Research Group, 2000 P Street, N.W., Suite 700, Washington, DC 20036.

From the Institute for Food and Development Policy, 145 9th Street, San Francisco, CA 94103. Include 15 percent ($1.00 minimum) for shipping:

What Can We Do? Food and Hunger: How You Can Make A Difference, by William Valentine and Frances Moore Lappe, Institute for Food and Development Policy, 1980, 70 pp. ($2.95)

Publications from the Center for Study of Responsive Law: These books, some of them already classics in their field, were written to inform, to involve, and to mobilize. They are designed to change one or more injustices and deficiencies through citizen action and awareness. A higher quality of daily democracy requires a higher quality of daily citizenship. These works can help guide the way toward that delightful commitment.

For the People, by Joanne Manning Anderson, with introduction by Ralph Nader, Addison-Wesley, 1977. A consumer handbook for community action, 379 pp. ($5.95)

The Madness Establishment, by Franklin D. Chu & Sharland Trotter Grossman Pub., 1974. An analysis of federal government programs on mental health centers, 232 pp. ($8.95 hard)

Congress Project Profile Kit – includes two profiles of past Members of Congress with 21-page bibliographic guideline for students to profile their current Member of Congress. ($2.50)

Company State, by James Phelan and Robert Pozen, with introduction by Ralph Nader, Grossman Pub., 1973. A study group report on Dupont in Delaware, 464 pp. ($3.95)

Politics of Land, Robert C. Fellmeth, with introduction by Ralph Nader (Grossman Pub., NY, 1973) – a study group report on land use in California, 715 pp. ($5.95)

How to Appraise and Improve Your Daily Newspaper, by David Bollier, with introduction by Ralph Nader, Disability Rights Center, 1980. A manual for readers. ($5.00)

Banding Together: How Check-offs Will Revolutionize the Consumer Movement, by Andrew Sharpless and Sarah Gallup, 1981. ($5.00)

Disposable Consumer Items: The Overlooked Mercury Pollution Problem, by John Abbotts, 1981. ($5.00)

Energy Conservation: A Campus Guidebook, by Kevin O'Brien and David Corn, 1981. ($5.00)

Women Take Charge: Asserting Your Rights in the Marketplace, by Nina Easton, preface by Ralph Nader, 1983, 202 pp. ($6.50)

Of Meters and Misfeasance: What You Should Know About Utility Metering and Billing Errors, by Brandon F. Greene and Show Mao Chen, 1986, 144 pp. ($12.00)

Being Beautiful, prepared by Katherine Isaac, with introduction by Ralph Nader, 1986. Selected readings on cosmetics safety and the beauty business, 246 pp. ($10.00)

Return To The Jungle, by Kathleen Hughes, with foreword by Ralph Nader, 1983. How the Reagan Administration is imperiling the nation's meat and poultry inspection program, 63 pp. ($6.50)

To order, make checks payable to:
Center for Study of Responsive Law
P.O. Box 19367
Washington, DC 20036

A new way to make the marketplace work for you.

Buyer's Market

Buyer's Market is Ralph Nader's consumer advisory, an eight-page monthly news letter packed full of useful techniques and tactics to improve your buying skills.

Volume One

1. Insurance
2. Food
3. Autos
4. Energy
5. Utilities
6. Health
7. Housing
8. Banking
9. Addictions
10. Cooperatives

Volume Two

1. Complaints
2. Credit
3. Professionals
4. Repairs
5. Bills
6. Travel
7. Women
8. Children
9. Diets
10. Shoes

Volume Three

1. Drinking Water
2. Cosmetics
3. Media
4. Food
5. Health
6. Indoor Pollution
7. Multinationals
8. Citizen Action
9. Toys
10. Television

A Message from Ralph Nader

Dear Friend,

    Are you eager to further improve your standard of living, save money and take on city hall or the corporate bureaucrats? *Buyer's Market* advisories will help you do just that. You can order Volume One, Volume Two and/or Volume Three.

    Issues contain:

- Concise information to help you stretch your hard-earned dollars;
- Advice straight from our experts - advice you may not get elsewhere;
- Strategies for consumer success that you can use every day; and
- Tips on bargain credit cards, travel plans, cosmetics and more.

    A Resources section on the back page of each advisory will offer free publications and other references should you wish to dig deeper into a topic and come out even further ahead in the marketplace. Being a savvy, confident buyer and citizen means better health, greater safety and more economic value for you and your family. If you have already subscribed for 1985, 1986 and 1987, please take this time to consider a gift subscription for a relative, friend or library.

                                  Sincerely,

                                  Ralph Nader

❑ I'd like to order my *Buyer's Market* volumes today! Please send me each volume (ten issues) for only $10.00 each:

❑ Volume 1 (1985) $_____

❑ Volume 2 (1986) $_____

❑ Volume 3 (1987) $_____

Total Enclosed $_____

Send to:

Name_____

Address_____

City _____ State _____ Zip _____

Make your check payable to:
Buyer's Market, P.O. Box 19367, Washington, D.C., 20036

For the latest news on foods and health, join the Center for Science in the Public Interest (CSPI) and receive *Nutrition Action Healthletter*. This is the newsletter relied upon by consumers and journalists from coast to coast, because the information is original, it's no-holds-barred honest, and it's understandable.

With billion-dollar advertising blitzes telling you how wonderful the all-new Tongue Teasers are, *Nutrition Action Healthletter* gives you the real poop. Ten times a year, this nationally acclaimed, 16-page, illustrated newsletter provides tasty, nourishing recipes, debunks deceptive ads, and gives you the lowdown on vitamin supplements.

Beyond giving you the information that will improve your diet and health and save you money, CSPI serves as your personal lobbyist and watchdog in Washington. We've stopped numerous deceptive advertising campaigns, gotten restrictions on unsafe food additives, and obtained improved food labeling. We hope you'll join 75,000 other concerned citizens and become part of our membership family.

Membership*

(#M1) One year: **$19.95** (**US$22.95** Canada & foreign)
(#MD) One year: **$14.95** (**US$17.95** Canada & foreign) for full-time students or senior citizens. (Back issues $2 ea.)

*Membership entitles you to 10 issues of **Nutrition Action Healthletter** a year, free "Nutrition Scoreboard" and "Chemical Cuisine" posters, and a 10 percent discount on all CSPI publications. All but $5 of your membership fee, which covers your subscription to **Nutrition Action Healthletter**, is tax-deductible.

Books and Pamphlets

Food Irradiation—Who Wants It?—Written by Tony Webb and prefaced by Michael Jacobson. Get the facts straight about one of this century's most controversial food preservation techniques. Price: **$5.95 (#FI)**.

The Changing American Diet—A graph-filled chronicle of national eating habits from 1910 to 1981. Price: **$4.00 (#47)**.

Creative Food Experiences for Children—A classic for teaching three- to ten-year-olds an awareness of good nutrition. Price: **$5.95 (#23)**.

The Complete Eater's Digest and Nutrition Scoreboard—Michael Jacobson's complete factbook of nutrition and food additives. Price: **$9.95 (#48)**.

Eat, Think, and Be Healthy!—A guide that's chock-full of enjoyable activities designed to teach nutrition to children ages 8–12. Price: **$8.95 (#24)**.

Fast Food Guide—The most complete and up-to-date compilation of nutrition and ingredient information for major chains. Price: **$4.95 (#FF)**.

Guess What's Coming to Dinner—CSPI's guidebook to contaminants in fresh meats and produce. Price: **$3.50 (#ASF)**

Marketing Booze to Blacks—An up-to-date report on the alcoholic beverage industry's campaign to increase drinking among black Americans. Price: **$4.95 (#66)**.

Salt: The Brand Name Guide to Sodium—Lists the sodium content of over 5,000 foods plus discussion of sodium and hypertension. Price: **$4.50 (#76)**.

Tainted Booze—Lists the urethane content (a carcinogen) of over 1,000 alcoholic beverages plus discussion. Price: **$3.95 (#65)**.

Posters

The following prices apply to all posters:

Standard: Quantities of 1 to 9, **$3.95** ea.; 10-49, **$2.75** ea.; 50-99, **$2.25** ea.; 100-499, **$1.25** ea.; 500+, *write for special prices.*

Laminated: Quantities of 1 to 9, **$7.95** ea.; 10-49, **$5.50** ea.; 50-99, **$4.00** ea.; 100-499, **$3.00** ea.; 500+, *write for special prices.*

Anti-Cancer Eating Guide—Which foods protect against cancer and which foods promote it? Here's how to fight cancer with your fork. (**Standard: #90; Laminated: #90L**).

Chemical Cuisine—Which food additives are safe and which are not? This poster has the answers. (**Standard: #80; Laminated: #80L**).

Exer-Guide—Learn how many calories you'll burn through exercise and the health benefits of keeping fit. (**Standard: #98; Laminated: #98L**).

Fast Food Eating Guide—Lists the fat, sodium, sugar, and calorie content of more than 200 fast food favorites. (**Standard: #94; Laminated: #94L**).

LIFE*SAVER Fat and Calorie Guide—Find out about the fat hidden in over 200 commonly eaten foods. (**Standard: #88; Laminated: #88L**).

New American Eating Guide—Dozens of foods are ranked as "Anytime," "In Moderation," and "Now and Then." (**Standard: #84; Laminated: #84L**).

Nutrition Scoreboard—How does your diet rate? Here over 200 foods are scored for their nutritional value. (**Standard: #82; Laminated: #82L**).

Sodium Scoreboard—Here's how to eat the low-sodium way. Sodium content of over 200 foods listed. (**Standard: #86; Laminated: #86L**).

Sugar Scoreboard—How sweet is it? The Scoreboard lists the sugar content of over 200 commonly eaten foods. (**Standard: #92; Laminated: #92L**).

Public Interest Software

Michael Jacobson's Nutrition Wizard—Analyze your diet, menus, recipes, and the nutritional content of hundreds of commonly available foods, beverages, and vitamin supplements. Price: **$99.95 (#D4)** *(IBM-compatible PCs only; 128K RAM).*

Order Form

Qty.		Membership Option	Price	Cost
	#M1	One-Year Membership	$19.95	
	#MD	Student/Senior Citizen Discount	$14.95	
A		MEMBERSHIP SUBTOTAL		$

Qty.	Prod. #	Name of Product	Price	Cost
		PRODUCTS SUBTOTAL		$
		MEMBERS: SUBTRACT 10% PRODUCT DISCOUNT		$
B		NEW PRODUCT SUBTOTAL		$
C	HANDLING	(Add $2.00 for expedited delivery and handling.)	$	.30
D		MEMBERSHIP SUBTOTAL (from A above)		$
		COST (B+C+D)		$
		Non-U.S. orders add 15% of cost for postage.		
		Tax-deductible contribution to CSPI's work (*Thanks!*)		
		NEW TOTAL COST		$

To order CSPI products or membership by phone, call us at **(202) 332-9110** Monday through Friday, 9 a.m. to 5 p.m. ET. Sorry, we cannot accept collect calls. Minimum order $19.95. Please have your credit card in hand when you call.

FULL MONEY BACK GUARANTEE...

Return any CSPI book or poster in 30 days for a full refund.

(CHECK ONE BOX)
☐ Payment enclosed (check or money order payable to "CSPI")
☐ Purchase orders from institutions only ($50 minimum order)
☐ Charge to credit card ($19.95 minimum order)   NAH-05
           ☐ VISA   ☐ MasterCard

_ _ _ _   _ _ _ _   _ _ _ _   _ _ _ _

_____   _____
Signature   Expiration date

Name   (PLEASE PRINT OR TYPE)

Address   Apt. Number

City   State   Zip

Send order form to: **CSPI-PD, 1501 16th St., NW Washington, DC 20036**

Non-U.S. Orders. Payment must be by Postal Money Order in U.S. Funds.

Please allow 4 to 6 weeks for delivery of publications; 6 to 8 weeks for your first issue of **Nutrition Action Healthletter**. For expedited delivery, add $2.00 in the space provided on the order form. Member discount does not apply to bulk purchases.